MATHEMATICS FOR ECOLOGISTS

I. CHASTON
B. Sc., Ph. D.

LONDON BUTTERWORTHS

THE BUTTERWORTH GROUP

ENGLAND
Butterworth & Co (Publishers) Ltd
London: 88 Kingsway, WC2B 6AB

AUSTRALIA
Butterworth & Co (Australia) Ltd
Sydney: 586 Pacific Highway Chatswood, NSW 2067
Melbourne: 343 Little Collins Street, 3000
Brisbane: 240 Queen Street, 4000

CANADA
Butterworth & Co (Canada) Ltd
Toronto: 14 Curity Avenue, 374

NEW ZEALAND
Butterworth & Co (New Zealand) Ltd
Wellington: 26–28 Waring Taylor Street, 1
Auckland: 35 High Street, 1

SOUTH AFRICA
Butterworth & Co (South Africa) (Pty) Ltd
Durban: 152–154 Gale Street

First published 1971

ISBN 0 408 70254 0 Standard
0 408 70255 9 Limp

Printed by photo-lithography and made in Great Britain at
the Pitman Press, Bath

PREFACE

In recent years the science of ecology has been revolutionised by the use of mathematics in data analysis. Descriptions of the complex equations which have been applied to ecological problems are now frequently appearing in the scientific literature, and, as many biologists do not have the opportunity of obtaining any formal undergraduate training in mathematics, they often have difficulty in fully understanding these new methods.

The purpose of this book, therefore, is to provide an introduction to some of the basic principles of advanced mathematics by illustrating how the basic concepts of calculus and linear algebra can be used to solve simplified biological problems.

It is hoped that this text will enable the reader to follow the underlying theory of ecological phenomena explained in terms of a mathematical model, and possibly stimulate him to study further the rapidly expanding field of mathematical ecology.

The author wishes to acknowledge the staff and students of the Department of Zoology at the University of Georgia for their stimulating discussions on the aspects of mathematical ecology during the period from September 1968 to June 1969; and the University of Newcastle upon Tyne for providing the facilities which permitted the development of many of the biological examples that form the basis of this text.

LONDON

I.C.

CONTENTS

One

CO-ORDINATES AND FUNCTIONS

1.1. Biological Example

A biologist working on the ecology of a freshwater insect examined the capability of the species to withstand the scouring effect of water current in an experimental stream. He found that, at a current speed of 5 cm/s, 5% of the animals in the stream were swept off the substrate and that, at 10 cm/s, the number losing their grip increased to 25%.

From these data, he then wished to estimate what proportion of the population would be swept into the water column in the field under spate conditions, when the water current was known to reach a speed of 20 cm/s.

To solve this problem, it is necessary to examine the geometrical theory of co-ordinate systems and straight lines.

1.2. Co-ordinates

The commonest co-ordinate form is the Cartesian, or rectangular, co-ordinate system, which is obtained by dividing a plane into four quadrants by the use of intersecting vertical and horizontal lines (Figure 1.1). These two lines are the axes of the system, with the horizontal line being taken as the X axis and the vertical line as the Y axis.

Any point P may be characterised by the co-ordinates (x, y), where x is the X co-ordinate (or abscissa) and y is the Y co-ordinate (or ordinate) of P. It is obvious (see Figure 1.1) that x is the distance of P from the Y axis and that y is the distance from the X axis. In addition, note that x has a negative value to the left of the Y axis and that y has a negative value below the X axis.

1.3. The Straight Line

A straight line is composed of a set of points whose co-ordinates (x, y) satisfy an equation of the type

$$ax + by + c = 0$$

An equation of this type is said to be the equation of the line.

If the equation is of the type

$$y = b$$

then the line will be parallel to the X axis and b units from it. If the equation is

$$x = a$$

then the line will be parallel to the Y axis and a units from this axis (Figure 1.2).

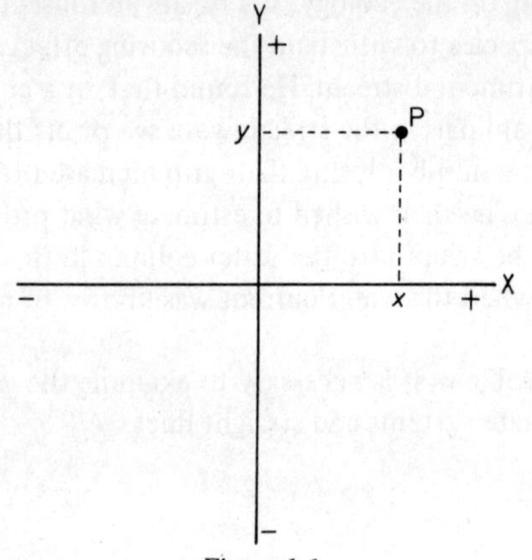

Figure 1.1

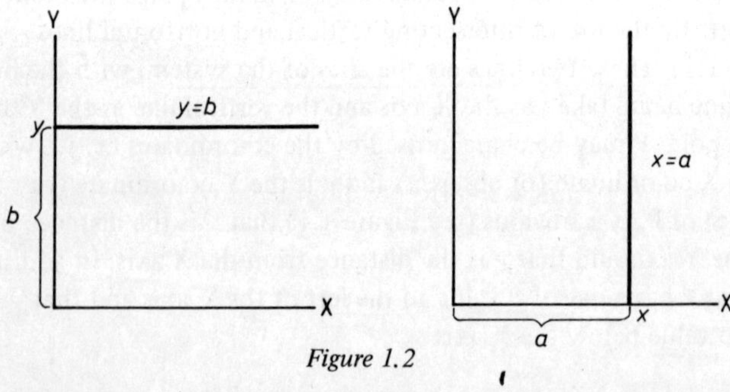

Figure 1.2

In Figure 1.3, the line L is defined by a point P through which L passes and by the angle α that L makes with the X axis. From trigonometry, it is known that the tangent of the angle α in the triangle P_2P_1Q is equal to the length of the side P_2Q divided by the length of the adjacent side P_1Q. As point P_2 is defined by the co-ordinates (x_2, y_2)

and point P_1 by the co-ordinates (x_1, y_1), the length of the side P_2Q is equal to $y_2 - y_1$ and that of the side P_1Q is equal to $x_2 - x_1$. In addition, if m is taken as the slope of the line L, then

$$m = \tan\alpha = \frac{P_2Q}{P_1Q} = \frac{y_2 - y_1}{x_2 - x_1} \tag{1.1}$$

This form of equation applies to any line which passes through points which are defined by Cartesian co-ordinates.

If $P(x, y)$ is any point on the line L, then the slope of the line segment P_2P (Figure 1.3) is the same as the slope of the segment P_1P_2 because the

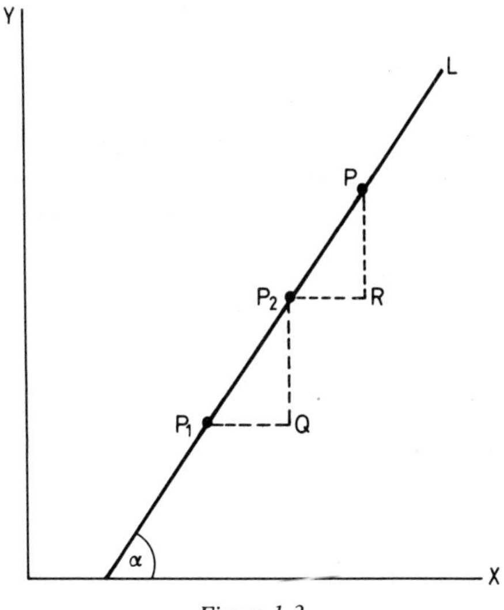

Figure 1.3

angles P_2P_1Q and PP_2R are corresponding angles of the parallel lines P_1Q and P_2R cut by the transversal L. Therefore

$$m = \tan\alpha = \frac{y_2 - y_1}{x_2 - x_1} = \frac{y - y_2}{x - x_2}$$

or

$$y - y_2 = \frac{y_2 - y_1}{x_2 - x_1}(x - x_2) \tag{1.2}$$

and thus, from equation 1.1,

$$y - y_2 = m(x - x_2) \tag{1.3}$$

1.4. Solution of the Biological Example

The concept of Cartesian co-ordinates can be applied to the problem of the proportion of the population swept into the water column under spate conditions (Section 1.1), by letting the X axis represent current speed and the Y axis represent the percentage of the population losing their grip of the substrate. The existing data can be plotted as $P_1(5, 5)$ and $P_2(10, 25)$, and, from equation 1.1, the slope of the line through these two points is

$$m = \frac{y_2 - y_1}{x_2 - x_1} = \frac{25 - 5}{10 - 5} = 4$$

Thus, for a unit increase in current speed of 1 cm/s, there is a 4% increase in the proportion of the animals losing their grip of the substrate.

The percentage being swept into the water column can be plotted as the point P_3 (Figure 1.4) with the co-ordinates $(20, y)$, where y is the unknown proportion which can be calculated from equation 1.3:

$$y - y_2 = m(x - x_2)$$

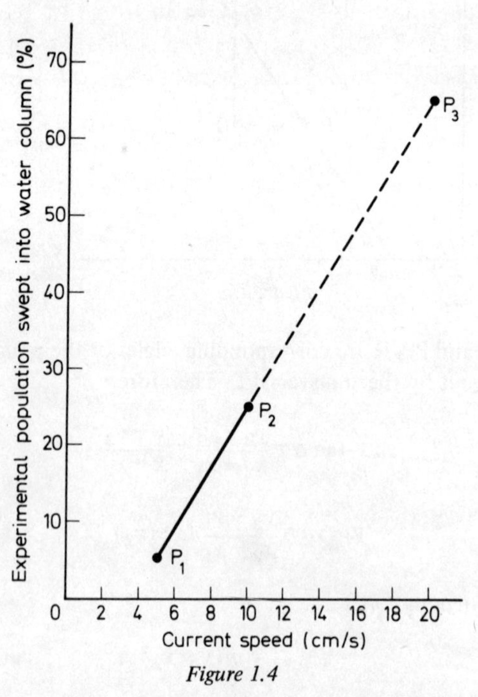

Figure 1.4

or
$$y = m(x - x_2) + y_2$$
$$= 4(20 - 10) + 25$$
$$= 65$$

Thus, from this calculation, the biologist can conclude that, under spate conditions, 65% of the population will be swept into the water column.

1.5. Polar Co-ordinates

With Cartesian co-ordinates, a point P is described in terms of its distance from the X and Y axes. Now it has been shown in Section 1.3 that a line is determined by the point through which it passes and the angle that it makes with the X axis. Thus one would expect that points on a line could be represented in terms of angular co-ordinates, and one method that utilises this type of point definition is the polar co-ordinate system.

A point P can be described in terms of polar co-ordinates as the length of the line joining P to the origin O and the angle that OP makes with a horizontal line Q (Figure 1.5). Thus P(r, θ) is the polar representation of a point P, where r is the length of OP and θ the angle that OP makes with Q.

The relationship between rectangular and polar co-ordinates can be found by letting the origin and the horizontal line of a polar co-ordinate system be at the same time the origin and an axis of a Cartesian system (Figure 1.6). If P is any point with the Cartesian co-ordinates (x, y) and the polar co-ordinates (r, θ), then, by the definition of sine and cosine,

$$\cos\theta = \frac{x}{r} \tag{1.4}$$

and

$$\sin\theta = \frac{y}{r} \tag{1.5}$$

The Cartesian co-ordinates (x, y) can then be described in terms of polar co-ordinates since, from equations 1.4 and 1.5,

$$x = r\cos\theta \tag{1.6}$$

and

$$y = r\sin\theta \tag{1.7}$$

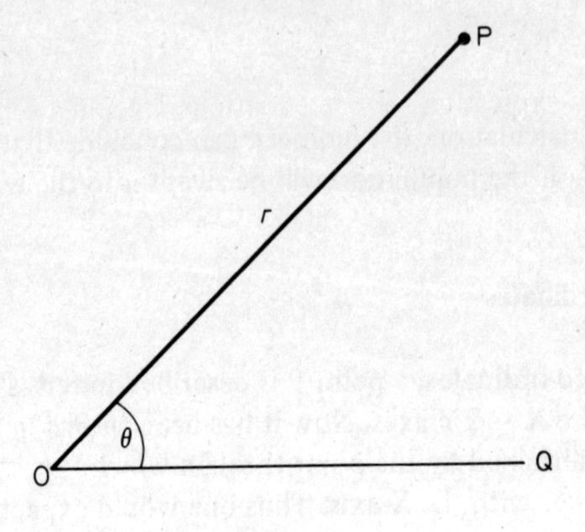

Figure 1.5

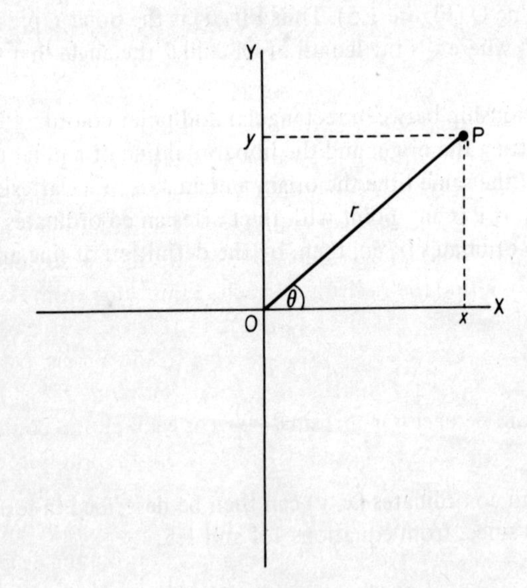

Figure 1.6

In addition, as r is the distance OP, then by the Pythagorean theorem concerning right-angled triangles

$$r = \sqrt{(x^2 + y^2)}$$

Substituting this expression for r in equations 1.4 and 1.5 gives

$$\cos\theta = \frac{x}{\sqrt{(x^2 + y^2)}} \quad (1.8)$$

and

$$\sin\theta = \frac{y}{\sqrt{(x^2 + y^2)}} \quad (1.9)$$

These two equations enable θ to be expressed in terms of Cartesian co-ordinates because

$$\theta = \arccos\left[\frac{x}{\sqrt{(x^2 + y^2)}}\right] \quad (1.10)$$

and

$$\theta = \arcsin\left[\frac{y}{\sqrt{(x^2 + y^2)}}\right] \quad (1.11)$$

1.6. The Biological Application of Polar Co-ordinate Systems

It is likely that, in plotting biological data, the reader will rarely utilise polar co-ordinate systems. One very useful application of it in ecology, however, is in the mapping of habitats.

For example, if one were mapping the location of clumps of bushes in an area in relation to defined boundaries X and Y using Cartesian co-ordinates, it would be necessary to take two measurements with a metre tape to define the position of each clump: one from the Y axis to the clump, along a line parallel to the X axis (Figure 1.7); and the second from the X axis, along a line parallel to the Y axis. On the other hand, if one used a polar co-ordinate system, only one measurement of distance would be necessary: the location of each clump could be described in terms of the distance of a line from the intersection of the boundaries to the clump, and the angle that this line makes with the X boundary (Figure 1.8). The measure of the angle can be achieved by placing a theodolite on the boundary intersection, taking a sighting along the X boundary and a second sighting on the clump, and then reading the angle recorded on the theodolite. Then this angle, plus the measurement taken with the metre tape from the intersection of the boundaries to the clump, gives the location of the clump. Consequently, one has halved the effort required for the often difficult task of accurate measurement of distance.

From equations 1.4 and 1.5, the Cartesian co-ordinates (x, y) can be described in terms of the polar co-ordinates.

1.7. Biological Example

A study of a grassland community was concentrated upon the activities of a specific herbivore which crops grass. The amount of energy in calories that this species obtains is related to the level of primary production in the area, the share of the total grass crop consumed by the

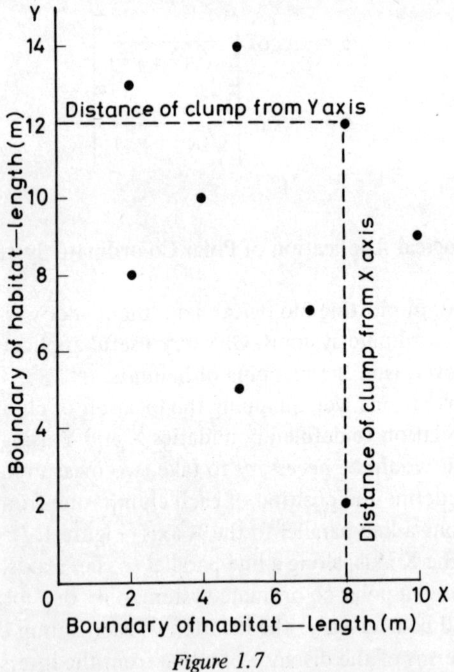

Figure 1.7

species in competition with other herbivores, and the energy exerted in obtaining food. From the raw data which have been collected, a biologist now wishes to describe quantitatively the return in energy that the herbivore species is obtaining from its feeding activities, and to use the developed model to theorise upon how the species could obtain even more energy from the grass crop.

To produce such a feeding strategy model, it is necessary to examine the relationships of the variables in the system, and this can only be achieved by understanding the properties of functions.

1.8. Functions

It was shown in Section 1.2 that any point in an XY plane can be represented by the co-ordinates (x, y); x and y are said to be the variables of the system because they are assigned real values in order to define a point P.

In the first biological example (Sections 1.1 and 1.4), the proportion of the population being swept into the water column under spate conditions was calculated by taking a value for current speed of 20 cm/s. In this case, therefore, the value for current speed determined the proportion

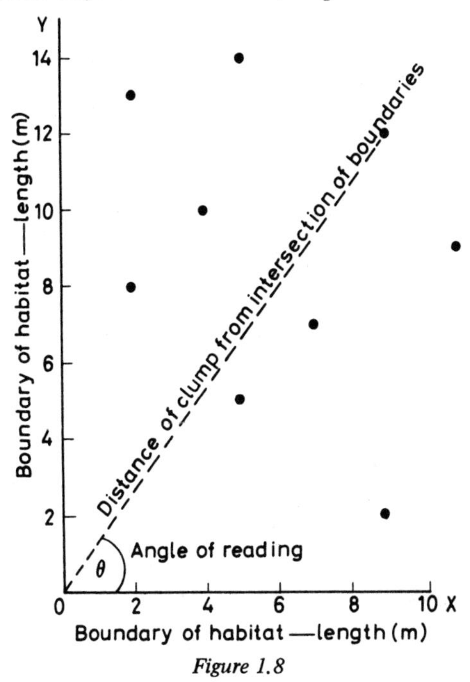

Figure 1.8

of the population losing their grip; or, more formally, the percentage of the population entering the water column was a 'function' of current speed. Thus, given a variable x which determines the value of y, it is said that x is the independent variable and y is the dependent variable because it is a function of x [denoted by $y = f(x)$].

1.9. The Graph of a Function

Although it is possible to describe a function in terms of an algebraic equation [for example, $y = f(x) = 2x + 1$] it is often more illuminating to plot a graph of the relationship. This can be achieved by constructing

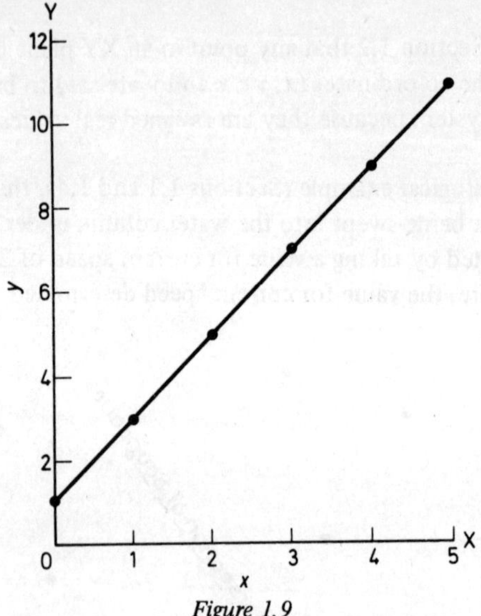

Figure 1.9

a table of values for the independent variable x and, for each of these values, calculating the respective value of the dependent variable y. It is then possible to use each tabulated value of x and the corresponding value for y as the co-ordinates of a Cartesian system and thus plot y against changing values of x.

For example, given the relationship

$$y = f(x) = 2x + 1$$

one can construct a table for values of x from 0 to 5 and calculate the corresponding values of y (Table 1.1). These values can then be plotted on a graph to illustrate the relationship between x and y (Figure 1.9).

Table 1.1

x	y
0	$(2 \times 0) + 1 = 1$
1	$(2 \times 1) + 1 = 3$
2	$(2 \times 2) + 1 = 5$
3	$(2 \times 3) + 1 = 7$
4	$(2 \times 4) + 1 = 9$
5	$(2 \times 5) + 1 = 11$

1.10. Solution of the Herbivore Strategy Problem

To produce a feeding strategy model (Section 1.7), it is necessary to separate the variables in the system, and these are as follows.

The relationship between total primary production and the grass crop. Primary production in any area is dependent upon such factors as the level of incident radiation and the availability of water and mineral compounds. In addition, grass production is one segment of the total primary production, and any change in the latter will be reflected in the

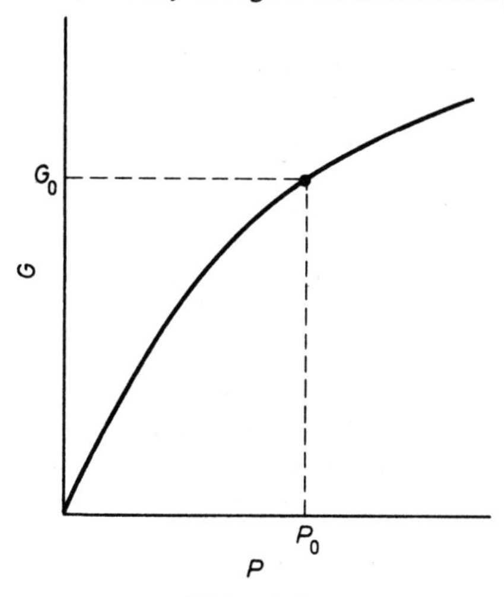

Figure 1.10

level of grass production. Therefore, grass production can be defined as a 'function' of the total primary production. Thus, if

P = total primary production (in calories), and
G = total grass production (in calories),

then

$$G = f_1(P) \tag{1.12}$$

and the relationship between the two variables can be shown graphically (Figure 1.10).

The relationship between total grass production and the share obtained by the herbivore species in competition with other species. Assuming the existence of interspecific competitive equilibrium, then, as the level of grass production increases, the herbivore species will obtain more grass. The share is therefore a function of total grass production. Thus, if

S = share of total grass production (in calories),

then

$$S = f_2(G) \tag{1.13}$$

and this relationship can be shown graphically (Figure 1.11).

(Note that the original relationships existing between the variables developed above are all denoted by the subscript figure 0 in their respective graphs.)

The relationship between the share of the grass crop and the amount of energy that the species obtains from its feeding activities. Food is the species' source of calories, and therefore the amount of energy obtained from the food is a function of the amount of grass consumed. Thus, if

E = energy from the food (in calories),

then

$$E = f_3(S) \tag{1.14}$$

and the relationship can also be shown graphically (Figure 1.12).

Although equation 1.14 describes the amount of energy in the food, the actual energy that the species obtains from its activities is equal to the total amount of calories in the food minus the cost in calories of the feeding activities. Thus, if

C = cost incurred in obtaining food (in calories),

then

$$E = S - C \tag{1.15}$$

The total cost C consists of two components: one is the maintenance metabolism of the species which exists even if the animals are not feeding; and the second is the cost of actually searching for the food, which can be taken as a fixed proportion of the total food obtained. Thus, if

M = maintenance metabolism cost, and
C_s = cost of searching for food, which in this example is $0{\cdot}1S$,

then, from equation 1.15,

$$\begin{aligned} E &= S - (M + C_s) \qquad (1.16) \\ &= S - (M + 0{\cdot}1S) \\ &= 0{\cdot}9S - M \end{aligned}$$

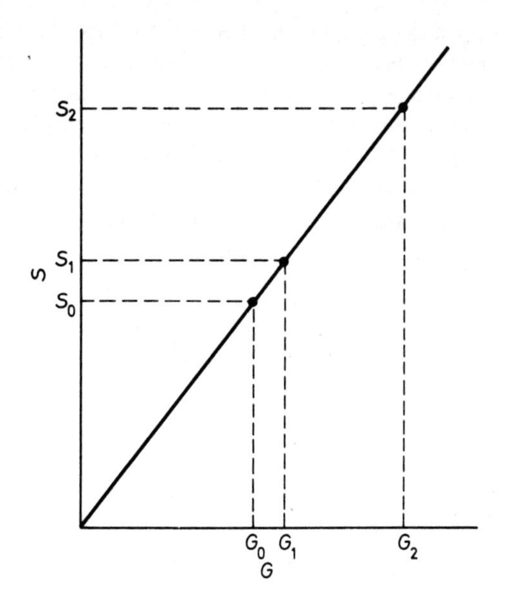

Figure 1.11

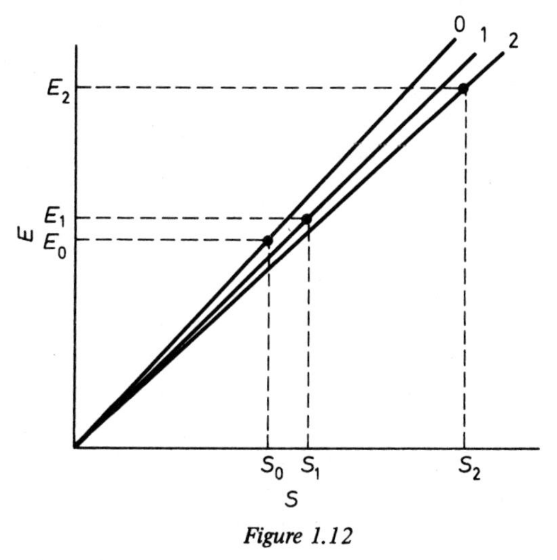

Figure 1.12

Having developed the basic model, it is now possible to examine strategies which would enable the herbivore to obtain more energy from the grass crop. To increase its share of the total crop, the species can either increase its searching afforts, or, on a much longer term basis, increase the size of the individual members of the species. However, such an evolutionary trend would cause an increase in the maintenance metabolism. Thus, if Δ denotes a defined amount of change, then, from equation 1.16, the new strategy is described by

$$E = S - [(M + \Delta M) + (C_s + \Delta C_s)]$$

$$= 0{\cdot}9S - \Delta C_s - (M + \Delta M) \qquad (1.17)$$

or, from equation 1.15,

$$E = S - (C + \Delta C) \qquad (1.18)$$

It is apparent from equation 1.18 that a new feeding strategy, by increasing the feeding costs, will decrease the amount of energy that the species obtains from its food, unless the increased costs cause a sufficient rise in the share of the grass crop obtained to compensate for the imbalance.

To demonstrate this effect, examine the situations in which:

a. The species increases its efforts to obtain food.
b. There is a selection pressure acting towards the evolution of larger individuals.

By referring to Figures 1.11 and 1.12, one can see the result of two such strategies. In case 1, although the increased costs result in an increased share of the grass crop, the actual energy obtained from the activities is approximately the same as in the original strategy. In case 2, however, the increased costs have resulted in such an increase in the share of the grass crop obtained that the species receives a much greater return in terms of energy than it did by operating its original strategy.

The above approach illustrates how one can use the simple concept of functions to produce a working biological model, and in fact this technique can be applied in formulating an ecological compartment model which can then be analysed on an analogue computer. In addition, it is interesting to note that the above model is a proof of the fact that the evolution of greater size is energetically disadvantageous unless it results in obtaining sufficient food to compensate for the increased maintenance metabolism costs. If the evolution of greater size in the species resulted in individuals obtaining less energetic profit from their feeding activities, there would no doubt be a selection pressure to remove the larger individuals of that species from the system. This selection pressure might not occur, however, if increased size conferred some other advantage such as a greater ability to defend themselves against predators.

Exercises

1. An animal at an age of 2 weeks is 4 cm long and, after another 4 weeks, its length has risen to 20 cm. Calculate the length of the animal when it is 10 weeks old.
2. Plot graphs of the functions $y = x^2$ and $y = 3x + 2$ for the values of x between 0 and 5.
3. By applying the concept of functions, develop a model to describe the relationships between the levels of a trophic pyramid by assuming that the number of animals at each level is dependent upon the number in the trophic level below it.

Two

DIFFERENTIAL CALCULUS

2.1. Biological Example

In any environment there is a limited amount of food available at each trophic level. Thus, if a species is present in very low numbers (and has no competitor for the same food), then it will not utilise all the available food and the total production for the species will be low. As the number of individuals of the species increases, the food supply will be utilised to a greater and greater extent until a point is reached where total production for the species is maximised. Past this point, the effects of intraspecific competition will become important because the number of individuals will be so high that a large amount of energy will be expended in competition and this factor will reduce the proportion of the food supply which can be utilised for production. It is therefore apparent that the level of total production for a species in an area is dependent upon the number of individuals of the species which are present in that area.

A biologist working on this phenomenon in freshwater environments carried out experiments to estimate the optimum number of fish that could be stocked in an area with a known food supply. For this research, he carried out parallel studies in which he stocked model streams with 3, 6, and 9 fish, and measured their increase in weight (in grams) after a known period. He repeated the experiments three times, and the calculated average weight increase for a single fish at the three levels of stocking were as shown in Table 2.1. This relationship between growth and population size can be illustrated by plotting the average increase in weight of a fish against stocking level (Figure 2.1).

From these data one can derive equations to express the growth in terms of the level of stocking. For, if

G = growth of single fish (in grams), and
N = number of fish stocked,

then, from equation 1.2,

$$G - y_2 = \frac{y_2 - y_1}{x_2 - x_1}(N - x_2)$$

Table 2.1

Total number of fish in stream	3	6	9
Average weight increase of a fish in a defined period (g)	350	200	50

Therefore, from the values in Figure 2.1,

$$G - 200 = \frac{200 - 350}{6 - 3}(N - 6)$$

$$= \frac{-150}{3}(N - 6)$$

$$= -50N + 300$$

and thus

$$G = 500 - 50N \tag{2.1}$$

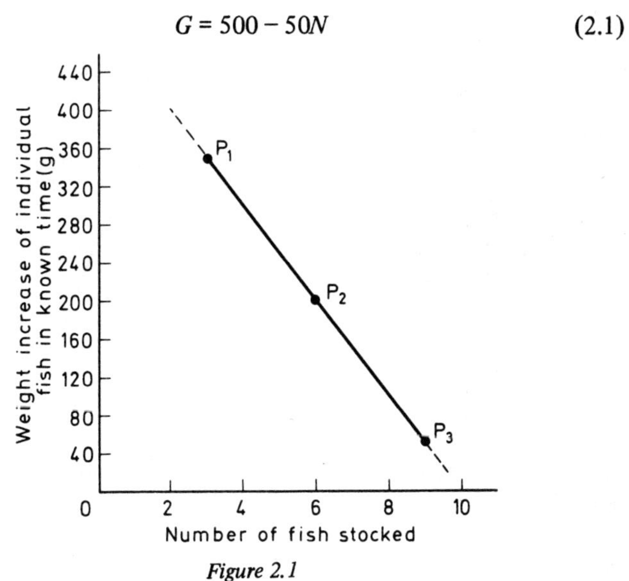

Figure 2.1

The total production at any level of stocking is related to the average growth of each fish at each stock level and to the number of fish stocked. Thus, if

T = total fish production (in grams),

then

$$T = GN$$

and thus, from equation 2.1,

$$T = (500 - 50N)N$$
$$= 500N - 50N^2 \qquad (2.2)$$

From equations 2.1 and 2.2, which describe the relationships between the variables, it is possible to examine the effects of changes in the independent variable upon the dependent variables. For example, if equation 2.2 is used to plot total production against the number of fish stocked, then it can be seen that the curve (Figure 2.2) reaches a maximum value when $N = 5$.

Although this graph illustrates the relationship between the variables, there is the drawback that the curve is obtained by plotting points

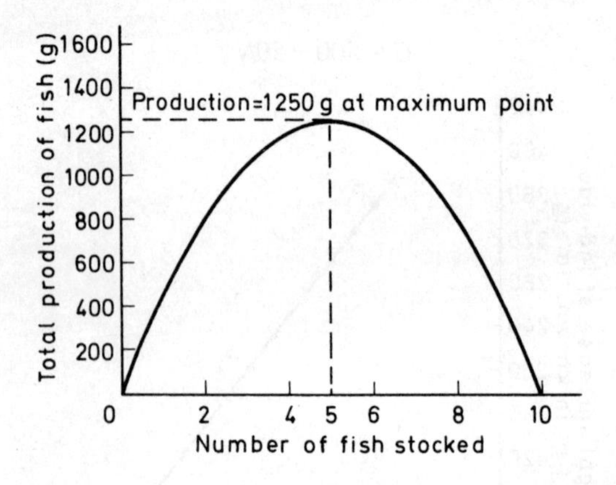

Figure 2.2

calculated using equation 2.2 and the accuracy of the graph is low unless one calculates a large number of points. This problem can be overcome, however, by using further equations to examine the relationship between the variables. For, if

ΔN = a change in N,
$N + \Delta N$ = an increase in the stock level,
T_0 = original total growth (in grams), and
T_1 = total growth after a change in stock level (in grams).

then, from equation 2.2,

$$T_0 = 500N - 50N^2$$

and $$T_1 = 500(N + \Delta N) - 50(N + \Delta N)^2$$

$$= 500N + 500\Delta N - 50N^2 - 100N(\Delta N) - 50(\Delta N)^2$$

Also, as

$$\Delta T = T_1 - T_0$$

then

$$\Delta T = 500N + 500\,\Delta N - 50N^2 - 100N(\Delta N) - 50(\Delta N)^2 - (500N - 50N^2)$$

$$= 500\Delta N - 100N(\Delta N) - 50(\Delta N)^2 \qquad (2.3)$$

To estimate the magnitude of the effect on T of changing the stock level N, one can divide ΔT by ΔN. Thus, from equation 2.3,

$$\frac{\Delta T}{\Delta N} = 500 - 100N - 50\Delta N \qquad (2.4)$$

As the change in N (i.e. ΔN) is made smaller and smaller (i.e. as ΔN approaches zero, denoted by $\Delta N \to 0$), then $\Delta T/\Delta N$ in equation 2.4 approaches the value $500 - 100N$, which is known as the limit of the equation.

The application of the limit concept to the ratio of change in the functional value per change in the value of the independent variable is the basis of differentiation in calculus, and in the above example $500 - 100N$ is called the derivative of the T function.

More formally, the limit of equation 2.4 is written

$$\underset{\Delta N \to 0}{\text{limit}} \frac{\Delta T}{\Delta N} = 500 - 100N$$

2.2. Differentiation

The concept of differentiation is the basis of differential calculus and, from the author's experience, one of the most difficult to convey to the biology student who has only a limited knowledge of mathematics. It is therefore worth dwelling upon the matter at some length.

Returning to the example in Sections 1.1 and 1.4, where the proportion of animals being swept into the water column was a function of current speed [$y = f(x)$], it was found that the expression

$$\frac{y_2 - y_1}{x_2 - x_1}$$

by describing the slope of a line between two points, was equivalent to the average change in percentage entering the water column per unit change in current speed.

Now, if d denotes any change in the value of the variables, then

$$\frac{dy}{dx} = \frac{y_2 - y_1}{x_2 - x_1} \tag{2.5}$$

For example, if

$$y = f(x) = x^2$$

then, from equation 2.5,

$$\frac{dy}{dx} = \frac{x_2^2 - x_1^2}{x_2 - x_1} \tag{2.6}$$

and, if $x_2 = x_1 + \Delta x_1$, where Δx_1 represents a small increment of current speed from x_1, then equation 2.6 can be rewritten as

$$\frac{dy}{dx} = \frac{(x_1 + \Delta x_1)^2 - x_1^2}{(x_1 + \Delta x_1) - x_1}$$

$$= \frac{x_1^2 + 2x_1(\Delta x_1) + (\Delta x_1)^2 - x_1^2}{\Delta x_1}$$

$$= 2x_1 + \Delta x_1$$

Reintroducing the concept of Δx_1 approaching zero, then, as this occurs, dy/dx will approach the limit $2x_1$. Or, more formally, for any value of x

$$\underset{\Delta x \to 0}{\text{limit}} \frac{dy}{dx} = 2x \tag{2.7}$$

As $\Delta x \to 0$, the expression dy/dx gets nearer and nearer the limit $2x$ at a speed x; in other words, equation 2.7 is expressing a rate of change of y with x at a single point and therefore can be said to be defining an instantaneous rate of change. Thus $2x$, the derivative of $y = f(x)$, is the instantaneous rate of percentage change of numbers entering the water column expressed as a function of the independent variable, current speed.

Note that, in this example, the result $dy/dx = 2x$ is a demonstration of the simple derivative law that, if

$$y = f(x) = x^n$$

where n indicates any power of x, then

$$\frac{dy}{dx} = nx^{n-1} \tag{2.8}$$

For example, if

$$y = f(x) = x^4$$

then

$$\frac{dy}{dx} = 4x^3$$

From the above discussion it is possible to define differentiation as a technique to find the instantaneous rate of change of a dependent variable in terms of the independent variable of which it is a function.

2.3. The Slope of a Curve

A more easily visualised definition of the result of differentiation is that the derivative at a point on a curve defines the slope of the curve at that point.

Consider a curve on which there are two points (x_1, y_1) and (x_2, y_2). Now the slope m of the line between these two points is given by

$$m = \frac{y_2 - y_1}{x_2 - x_1}$$

Now if

$$y = f(x) = x^2$$

then the slope is given by

$$m = \frac{x_2^2 - x_1^2}{x_2 - x_1} \tag{2.9}$$

and if $x_2 = x_1 + \Delta x_1$, then equation 2.9 is equivalent to

$$m = \frac{x_1^2 + 2x_1 (\Delta x_1) + (\Delta x_1)^2 - x_1^2}{x_1 + \Delta x_1 - x_1}$$

$$= 2x_1 + \Delta x_1 \tag{2.10}$$

If the value of Δx_1 approaches zero, then the point $x_1 + \Delta x_1$ will approach the point x_1. As this occurs, the slope of the line between the two points will approach the derivative $2x_1$, which can therefore be said to define the slope of the curve at the point (x_1, y_1).

Having shown that the derivative defines the slope of the curve at a point on a graph, it can also be concluded from the earlier discussion of differentiation that the slope of a line at a point on a graph expresses the instantaneous rate of change of the dependent variable at that point on the graph in terms of the independent variable of which it is a

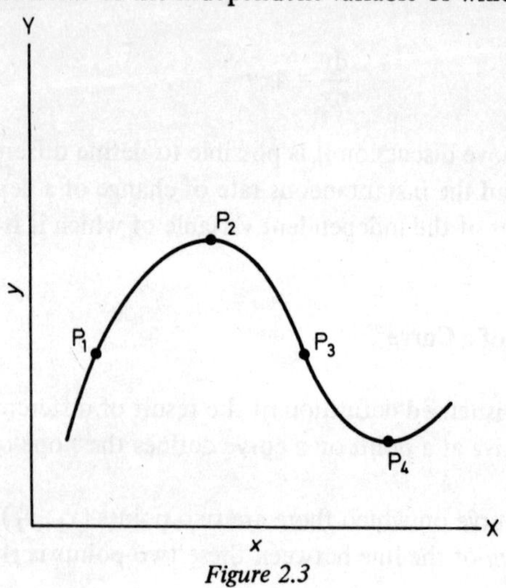

Figure 2.3

function. Thus consider the graph in Figure 2.3 and the four given points. At point P_1, y is increasing as x increases, or

$$\frac{dy}{dx} > 0 \text{ (i.e. positive)}$$

At point P_2, y is stationary, or

$$\frac{dy}{dx} = 0$$

At point P_3, y is decreasing as x increases, or

$$\frac{dy}{dx} < 0 \text{ (i.e. negative)}$$

At point P_4, y is stationary, or

$$\frac{dy}{dx} = 0$$

2.4. Notation

One may meet a number of different notations in differentiation, and it should be realised that, where $y = \mathrm{f}(x)$,

$$\frac{\mathrm{d}y}{\mathrm{d}x} \equiv \frac{\mathrm{d}\,[\mathrm{f}(x)]}{\mathrm{d}x} \equiv \frac{\mathrm{d}}{\mathrm{d}x}\,\mathrm{f}(x)$$

and that

$$\frac{\mathrm{d}y}{\mathrm{d}x} \equiv \mathrm{f}'(x) \equiv y'$$

2.5. Some Standard Derivatives

2.5.1. $y = \mathrm{f}(x) = x^n$

If

$$y = \mathrm{f}(x) = x^n$$

and n is any number, then

$$\frac{\mathrm{d}y}{\mathrm{d}x} = nx^{n-1} \tag{2.11}$$

2.5.2. $y = \mathrm{f}(x) = c$

If

$$y = \mathrm{f}(x) = c$$

and c is a constant, then, for all x,

$$\frac{\mathrm{f}(x + \Delta x) - \mathrm{f}(x)}{x + \Delta x - x} = \frac{c - c}{\Delta x}$$

Thus

$$\frac{\mathrm{d}y}{\mathrm{d}x} = 0 \tag{2.12}$$

i.e. the derivative of a constant is zero because $y = c$ is a line with zero slope.

2.5.3. $y = c\mathrm{f}(x)$

If

$$y = c\mathrm{f}(x)$$

and c is a constant, then, for all x,

$$\frac{c\mathrm{f}(x + \Delta x) - c\mathrm{f}(x)}{x + \Delta x - x} = \frac{c\,[\mathrm{f}(x + x) - \mathrm{f}(x)]}{\Delta x}$$

and thus

$$\frac{\mathrm{d}y}{\mathrm{d}x} = c\mathrm{f}'(x) \tag{2.13}$$

For example, if

$$y = 3\mathrm{f}(x) = 3x^2$$

then

$$\begin{aligned}\frac{\mathrm{d}y}{\mathrm{d}x} &= 3\mathrm{f}'(x)\\ &= 3(2x)\\ &= 6x\end{aligned}$$

2.5.4. $y = \mathrm{f}(x) = \mathrm{g}(x) + \mathrm{h}(x)$

If

$$y = \mathrm{f}(x) = \mathrm{g}(x) + \mathrm{h}(x)$$

then

$$\mathrm{f}'(x) = \mathrm{g}'(x) + \mathrm{h}'(x) \tag{2.14}$$

For example, if

$$y = \mathrm{f}(x) = x^4 + x^2$$

then

$$\mathrm{f}'(x) = 4x^3 + 2x$$

2.5.5. $y = \mathrm{f}(x) = \mathrm{g}(x)\,\mathrm{h}(x)$

If

$$y = \mathrm{f}(x) = \mathrm{g}(x)\,\mathrm{h}(x)$$

then

$$\mathrm{f}'(x) = \mathrm{g}'(x)\,\mathrm{h}(x) + \mathrm{g}(x)\,\mathrm{h}'(x) \tag{2.15}$$

For example, if

$$y = \mathrm{f}(x) = (3x^2 + 2x)\,(x^3)$$

then

$$\begin{aligned}\mathrm{f}'(x) &= (6x + 2)\,(x^3) + (3x^2 + 2x)\,(3x^2)\\ &= (6x^4 + 2x^3) + (9x^4 + 6x^3)\\ &= 15x^4 + 8x^3\end{aligned}$$

2.5.6. $y = \mathrm{f}(x) = \mathrm{g}(x)/\mathrm{h}(x)$

If

$$y = \mathrm{f}(x) = \frac{\mathrm{g}(x)}{\mathrm{h}(x)}$$

then

$$\mathrm{f}'(x) = \frac{\mathrm{g}'(x)\,\mathrm{h}(x) - \mathrm{g}(x)\,\mathrm{h}'(x)}{[\mathrm{h}(x)]^2} \tag{2.16}$$

For example, if

$$y = \mathrm{f}(x) = \frac{3x^2 + 2}{x^2}$$

then

$$\begin{aligned}\mathrm{f}'(x) &= \frac{(6x)\,(x^2) - (3x^2 + 2)\,(2x)}{(x^2)^2}\\ &= \frac{6x^3 - 6x^3 - 4x}{x^4}\\ &= \frac{-4}{x^3}\end{aligned}$$

2.5.7. The Chain Rule

Consider the function $y = f(z)$, where $z = g(x)$. It can be said that y is a compound function of x such that

$$y = F(x) = f\,[g(x)]$$

Then

$$F'(x) = f'(z)\,g'(x)$$

or

$$\frac{dy}{dx} = \frac{dy}{dz}\,\frac{dz}{dx} \tag{2.17}$$

For example, if

$$y = (x^2 + x)^4$$

and if

$$z = x^2 + x$$

then

$$y = z^4$$

Thus

$$\frac{dy}{dx} = \frac{dy}{dz}\,\frac{dz}{dx}$$

$$= 4z^3\,(2x + 1)$$

$$= 4(x^2 + x)^3\,(2x + 1)$$

2.5.8. Other Common Derivatives

If

$$y = f(x) = \exp x$$

where exp denotes the Napierian or natural logarithm, then

$$f'(x) = \exp x$$

If

$$y = f(x) = \exp g(x)$$

Then

$$f'(x) = g'(x) \exp g(x)$$

For example, if

$$y = f(x) = \exp x^2$$

then

$$f'(x) = 2x \exp x^2$$

If

$$y = f(x) = \log_e x$$

then

$$f'(x) = \frac{1}{x}$$

If

$$y = f(x) = \sin x$$

then

$$f'(x) = \cos x$$

If

$$y = f(x) = \cos x$$

then

$$f'(x) = -\sin x$$

2.6. Higher Derivatives

If one has found a derivative f′ of a function and the derivative itself is differentiable, then its derivative is denoted by f″

For example, if

$$f(x) = x^3 + x^2$$

then

$$f'(x) = 3x^2 + 2x$$

which is the first-order derivative; and

$$f''(x) = 6x + 2$$

which is the second-order derivative; and

$$f'''(x) = 6$$

which is the third-order derivative. Thus one can continue to take higher derivatives.

In the above example, the third-order derivative $f'''(x)$ is the highest derivative not equal to zero; differentiation of the constant 6 will give a fourth-order derivative equal to zero.

2.7. The Maximisation of Fish Production Problem

Returning to this earlier problem (Section 2.1), one can now examine the relationship between the rate of change of total production T of the fish population and changes in the number of fish stocked N by differentiating the equation

$$T = 500N - 50N^2$$

with respect to N.

Thus

$$\frac{dT}{dN} = 500 - 100N \qquad (2.18)$$

In other words, an increase in the number of fish stocked will cause an increase in the total production until the point where

$$500 = 100N$$

At this point

$$\frac{dT}{dN} = 0$$

and after this point

$$\frac{dT}{dN} < 0$$

because $100N > 500$; thus any increase in stock level will then decrease the total production of the fish population.

Summarising:

1. When $dT/dN > 0$ ($100N < 500$), any increase in stock level would decrease the average growth of each fish but increase the total production of the population.

2. When $dT/dN = 0$ ($100N = 500$), then the total production is at the maximum point.
3. When $dT/dN < 0$ ($100N > 500$), then the average growth of each fish will continue to decline if stock levels are increased and this will decrease the total number of grams of production.

2.8. The Lotka–Volterra Equations

Some of the equations which form the fundamental theories of ecology are those developed by Lotka and Volterra to describe: (a) the growth of a population in a limited environment; (b) interspecific competition; and (c) predation.

All these equations use differential calculus to describe rates of change in population size in relation to time. For example, the equation for population growth in a limited environment is

$$\frac{dN}{dt} = \frac{rN(K-N)}{K} \tag{2.19}$$

where

N = number of animals,
t = time,
r = rate of increase (i.e. birth rate minus death rate), and
K = 'carrying capacity' of the environment (i.e. the maximum possible size of the population that can exist in the area).

Equation 2.19 describes the type of rate change in population size shown in Figure 2.4, where the number of individuals N is plotted against time t.

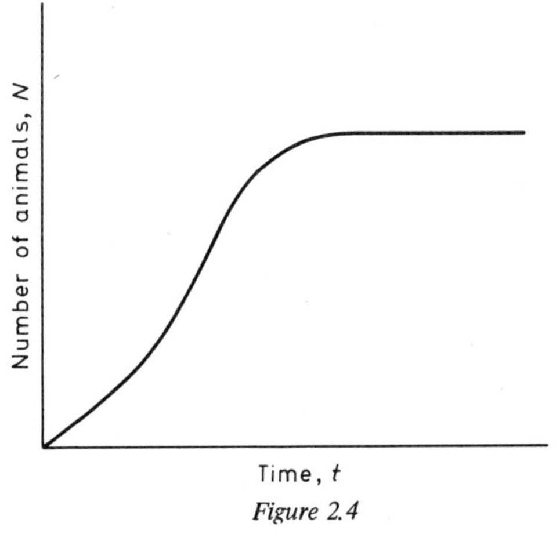

Figure 2.4

Although equation 2.19 is in fact a differential equation (see Chapter 6 for details on this type of equation) and not just a simple derivative, it can still be examined to illustrate how rate changes are applicable as descriptions of ecological phenomena.

For example, suppose there is a population where $K = 100$ and $r = +1{\cdot}5$ (i.e. birth rate exceeds death rate). Then, from equation 2.19, it is possible to calculate the rate of increase in animal numbers (dN/dt) at various levels of population size N, as shown in Table 2.2. It can be

Table 2.2

N*	$rN(K-N)/K$	dN/dt
10	$1{\cdot}5 \times 10(100-10)/100$	13·5
20	$1{\cdot}5 \times 20(100-20)/100$	24·0
40	$1{\cdot}5 \times 40(100-40)/100$	36·0
50	$1{\cdot}5 \times 50(100-50)/100$	37·5
60	$1{\cdot}5 \times 60(100-60)/100$	36·0
80	$1{\cdot}5 \times 80(100-80)/100$	24·0
100	$1{\cdot}5 \times 100(100-100)/100$	0·0

* Note that the number present N is a function of time.

seen from the table that the population growth rate dN/dt increases as N increases until it reaches a maximum when $N = 50$. This maximum point is known as the inflection point of the curve, because the steepness of the graph begins to decrease after this point as N gets nearer and nearer to K.

Eventually, the number of animals equals the carrying capacity of the environment (i.e. $N = K = 100$) and at this point the curve flattens to become a straight line, the rate of increase then being zero, i.e.

$$\frac{dN}{dt} = \frac{1{\cdot}5 \times 100\,(100-100)}{100}$$

$$= 0$$

The equations describing interspecific competition and predation are based upon modifications of equation 2.19. For interspecific competition,

$$\frac{dN_1}{dt} = \frac{r_1 N_1 (K_1 - N_1 - \alpha N_2)}{K_1} \qquad (2.20)$$

where

N_1 = number of members of species 1,
N_2 = number of members of species 2 (the competitor), and
α = degree of effect of N_2 on N_1.

For predation,

$$\frac{dN_1}{dt} = \frac{r_1 N_1 (K_1 - N_1 - \beta P)}{K_1} \tag{2.21}$$

where

P = number of predators, and
β = predation rate.

In both equation 2.20 and equation 2.21, it is apparent that another factor is acting (i.e. αN_2 or βP) over and above those in equation 2.19. The extra factor will cause the expression on the right-hand side of the equation to approach zero at a faster rate. Consequently, the population level of N_1 will usually be lower at maximum under predation or interspecific competition than the level attained in equation 2.19 for a population in a limited environment.

The actual population level reached in systems described by equations 2.20 and 2.21 is dependent upon the magnitude of competition (αN_2) or predation (βP), and it is possible that in some cases these effects will be so strong as to cause extinction of N_1.

Exercises

Differentiate the following functions:

1. $f(x) = x^2$
2. $f(x) = x^3$
3. $f(x) = 2x^5$
4. $f(x) = 2$
5. $f(x) = 3x + x^3$
6. $f(x) = x^2(2x)$
7. $f(x) = (4x - 3)/x^2$
8. $f(x) = \exp x$
9. $f(x) = \exp 2x$
10. $f(x) = \exp x^3$

Three

MAXIMUM AND MINIMUM POINTS

3.1. Maxima and Minima

At the end of the previous chapter, the derivative was used to find the maximum total production of a fish population. This was achieved by examining the derivative T' on each side of the value $\mathrm{d}T/\mathrm{d}N = 0$. It would not have been sufficient just to state that the total production was maximised when $\mathrm{d}T/\mathrm{d}N = 0$, because had the relationship between total production and level of stocking been of the type illustrated in Figure 2.3, $\mathrm{d}T/\mathrm{d}N = 0$ would occur at two points.

Thus all one can say of a point on a curve where $\mathrm{d}T/\mathrm{d}N = 0$ is that it is an extreme point. It is possible, as in Chapter 2, to evaluate whether the extreme point is a maximum or minimum point by looking at values on each side of the point, but such a task is laborious. It is simpler to use the properties of the second derivative to determine the nature of the extreme point.

A continuous function $y = \mathrm{f}(x)$ is said to have an extreme point if $y' = 0$. If a second derivative exists, then it is possible to evaluate the extreme point: when $y'' < 0$ (i.e. negative), the point is a maximum point, and, when $y'' > 0$ (i.e. positive), the point is a minimum point.

These statements of the conditions necessary for maxima and minima can probably be clarified by an example. If

$$y = 5 + 3x - x^3 \tag{3.1}$$

then

$$y' = 3 - 3x^2$$

The function has an extreme point when

$$y' = 0$$

and this occurs when

$$3 - 3x^2 = 0$$

i.e. when

$$x = \pm 1$$

Now

$$y'' = -6x$$

and from the above criteria for evaluating extreme points, it is known that, when $y'' < 0$, the point is a maximum point and, when $y'' > 0$, it is a minimum one. Thus, when $x = +1$,

$$y'' = -6 \text{ (i.e. } y'' < 0)$$

and the point is a maximum point. When $x = -1$,

$$y'' = +6 \text{ (i.e. } y'' > 0)$$

and the point is a minimum point.

The maximum and minimum values of y can now be obtained by replacing x by the value ± 1 in the original equation 3.1. When $x = +1$,

$$\begin{aligned} y &= 5 + 3 - 1 \\ &= 7 \text{ (a maximum point)} \end{aligned}$$

and, when $x = -1$,

$$\begin{aligned} y &= 5 - 3 + 1 \\ &= 3 \text{ (a minimum point)} \end{aligned}$$

Note that these points are only *local* maximum and minimum values because it is possible that other points may exist where y achieves values of more than 7 or less than 3. They would only be absolute points if the range of values for x is restricted, for example between the values $x = -2$ and $x = +2$.

3.2. Biological Example

A biologist working on the population dynamics of a planktonic crustacean found that, in laboratory experiments, as the number of adults placed in a culture of a defined volume was increased, the number of eggs per female decreased. From these experiments he produced a graph (Figure 3.1) and wished to calculate the maximum number of eggs that could be produced by the population in the culture experiments.

The same approach is used as in Section 2.1 to calculate the relationship between the number of eggs and population size, i.e.

$$E - y_2 = \frac{y_2 - y_1}{x_2 - x_1}(F - x_2)$$

where

E = number of eggs per female, and
F = number of females in the culture.

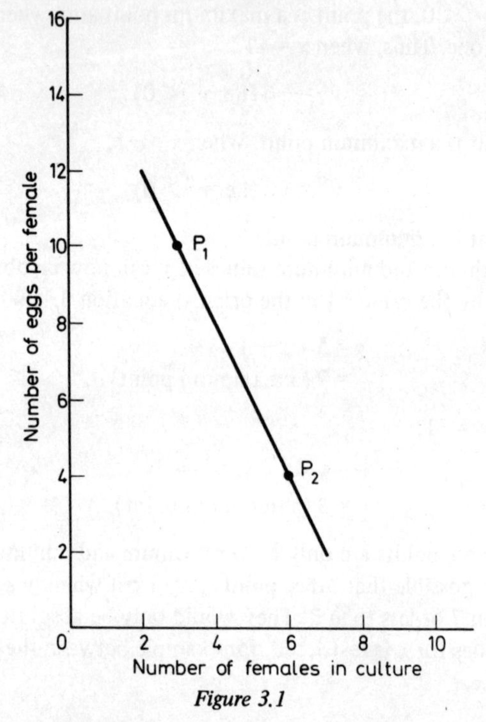

Figure 3.1

Thus, from the data in Figure 3.1,

$$E - 4 = \frac{4 - 10}{6 - 3}(F - 6)$$

or

$$E = 16 - 2F$$

Thus the total number of eggs T produced by the population is given by

$$T = EF$$
$$= 16F - 2F^2$$

Now, during the experiments, the biologist noted that the number of eggs which actually hatched was only 70% of the total number produced. This fact he described in the equation

$$H = T - D \tag{3.2}$$

where

H = number hatching, and
D = number of deaths (i.e. 100% − 70% = 30% of the total produced).

Consequently the instantaneous rate of hatching in relation to population size is given by

$$\frac{dH}{dF} = \frac{dT}{dF} - \frac{dD}{dF}$$

and the extreme point of this equation occurs when

$$\frac{dH}{dF} = \frac{dT}{dF} - \frac{dD}{dF} = 0$$

i.e. when

$$\frac{dT}{dF} = \frac{dD}{dF}$$

Having defined the extreme point, it is possible to evaluate this point from the second derivative

$$H'' = T'' - D''$$

The extreme point is a maximum point when

$$H'' < 0$$

i.e. when

$$T'' < D''$$

Thus, from equation 3.2, it is possible to estimate the maximum possible number of eggs hatching because

$$\begin{aligned} H &= T - 0{\cdot}3T \\ &= 16F - 2F^2 - 4{\cdot}8F - 0{\cdot}6F^2 \\ &= 11{\cdot}2F - 1{\cdot}4F^2 \end{aligned} \tag{3.3}$$

and thus

$$H' = 11{\cdot}2 - 2{\cdot}8F \tag{3.4}$$

An extreme point occurs when $H' = 0$ or

$$F = \frac{11{\cdot}2}{2{\cdot}8} = 4$$

and it is a maximum point because

$$H'' = -2{\cdot}8 \text{ (i.e. } H'' < 0)$$

The biologist now knows that the maximum number of total hatchings which can occur in the culture will be at the point where the number of females $F = 4$, because at this point $H' = 0$ and $H'' < 0$. From this information he can then calculate the maximum total of hatchings, for at $F = 4$

$$H = (11{\cdot}2 \times 4) - (1{\cdot}4 \times 16)$$

which (taken to the nearest whole number) gives 22 hatchings.

3.3. A Further Application of Maxima and Minima

Assume that a seed-eating bird has to supply its young with a fixed total amount of food F_T during the time that the offspring remain in the nest. To obtain this food, the parent must leave the nest to find the seeds and then transport them back to the young.

In its search it is assumed that the seeds are found on randomly distributed clumps of bushes and that there are more seeds in a single clump than the bird needs to satisfy totally its young. The actual cost of obtaining the food is equal to a fixed cost M, which represents the maintenance metabolism of the parent, plus a food finding cost $C_1 N$, where N is the number of seeds collected and C_1 is the cost of obtaining one seed. This latter value C_1 is a cost which decreases as more seeds are collected.

The decrease in C_1 is a result of the fact that the bird has the relatively large cost of finding the clump of bushes (and this cost is approximately the same for each trip because the clumps are randomly distributed), but the collection of each seed at the feeding site only involves a small cost. Thus the more the animal collects, the lower and lower is the unit cost, i.e.

$$\text{Cost per seed collected} = \frac{\text{Search cost + Cost at feeding site}}{\text{Number of seeds collected}}$$

This effect can probably be more clearly demonstrated by assuming real values in the equation to calculate the cost per seed. Thus, if the search cost is 20 cal and the cost per seed collected at the feeding site is 1 cal, then the cost per seed unit is as shown in Table 3.1.

Table 3.1

Number of seeds	*Search cost* (cal)	*Feeding site cost* (cal)	*Cost per seed unit* (cal)
1	20	1	21
2	20	2	11
3	20	3	7·6
4	20	4	6·0
5	20	5	5·0

The total collecting cost during the season is equal to the cost of obtaining food $(M + C_1N)$ multiplied by the number of trips that the bird makes. This latter value is equal to the total food F_T required by the young during the season, divided by the number of seeds collected in a single trip. Thus, if

C_c = total collecting cost (in calories),

Then

$$C_c = \frac{F_T}{N}(M + C_1N) \tag{3.5}$$

The other cost to be considered is that involved in transporting the collected food back to the nest: the carrying cost per seed C_2 is dependent upon the distance D covered, multiplied by the cost c of carrying one seed for one unit of distance, i.e.

$$C_2 = Dc \tag{3.6}$$

The total transporting cost C_t is equal to the carrying cost per seed unit, multiplied by the total number of seeds carried, i.e.

$$C_t = C_2N \tag{3.7}$$

Thus the total energy E_T expended in obtaining food during the season and transporting the collected seeds back to the nest can be described by the equation

$$E_T = \frac{F_T}{N}(M + C_1N) + C_2N$$

$$= \frac{F_TM}{N} + \frac{F_TC_1N}{N} + C_2N \tag{3.8}$$

If the collecting and transporting costs are plotted (Figure 3.2), then the total energy E_T for any seed amount can be obtained by summing the two component costs. From this operation it is apparent that E_T initially decreases as N increases and then, at a certain value of N, E_T begins to increase. Therefore it can be concluded that there is an

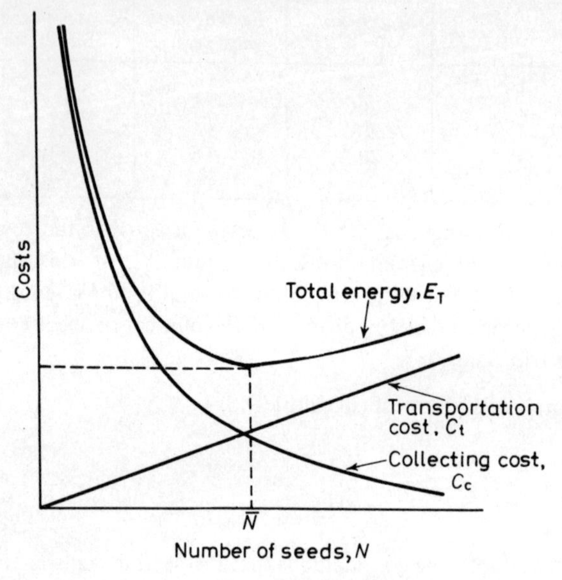

Figure 3.2

optimum number of seeds that the bird should collect in any one trip in order to minimise the total energy expended in feeding its young.

To calculate the optimum seed number $\bar{N}$, one simply differentiates E_T with respect to N and then examines the value of N when E'_T equals zero. Thus, from equation 3.8,

$$\frac{dE_T}{dN} = \frac{(F_T M)'N - (F_T M)N'}{N^2} + \frac{(F_T C_1 N)'N - (F_T C_1 N)N'}{N^2} + (C_2 N)'$$

(The reader who is confused by this operation is referred back to Section 2.5.6.). And since, if $y = f(x) = x$, $f'(x) = 1$, then

$$\frac{dE_T}{dN} = \frac{(0)N - (F_T M)}{N^2} + \frac{(F_T C_1)N - (F_T C_1 N)}{N^2} + C_2$$

$$= \frac{-F_T M}{N^2} + C_2 \qquad (3.9)$$

An extreme value of E_T occurs when $dE_T/dN = 0$, which happens when

$$\frac{F_T M}{N^2} = C_2$$

or when

$$N^2 = \frac{F_T M}{C_2}$$

i.e. when

$$N = \sqrt{\frac{F_T M}{C_2}}$$

Thus an extreme point exists when $N = \sqrt{(F_T M/C_2)}$, and in order to find whether this is a minimum point it is necessary to examine the second derivative. From equation 3.9,

$$\frac{d^2E_T}{dN^2} = \frac{-[(F_T M)' N^2 - (F_T M)(N^2)']}{N^4} + (C_2)'$$

$$= \frac{-[(0) N^2 - (F_T M) 2N]}{N^4} + 0$$

$$= \frac{+2F_T M}{N^3}$$

which is > 0, and therefore the extreme point given by $dE_T/dN = 0$ is a minimum point. From this it can be concluded that the bird is operating an optimum strategy of minimum energy expenditure when it collects the number of seeds $\bar{N}$ given by

$$\bar{N} = \sqrt{\frac{F_T M}{C_2}}$$

Four

FUNCTIONS INVOLVING MORE THAN ONE VARIABLE

4.1. Biological Example

A study of a bird showed that the number of prey it captured N during a certain period could be described by the equation

$$N = 100 + 2S + 0{\cdot}5ST + T - 10P \tag{4.1}$$

where

S = energy expended during the search for prey in the feeding area (in calories)
T = total energy expended in travelling from its roost to a feeding area (in calories), and
P = unit value of the prey (in calories).

Equation 4.1 describes the effect that, as more energy is expended by the bird in travelling to a suitable feeding area and in searching for prey, then the greater will be the number of prey captured. This situation is modified, however, by the calorific value of the prey units, because, the greater the unit value of the prey, then the fewer is the number of prey captured, i.e. if the unit value of the prey is high, the bird will soon become satiated and thus cease to feed. Now if

F = total calorific value of all the prey consumed,

then

$$F = PN$$

or, from equation 4.1

$$\begin{aligned} F &= P(100 + 2S + 0{\cdot}5ST + T - 10P) \\ &= 100P + 2SP + 0{\cdot}5STP + TP - 10P^2 \end{aligned} \tag{4.2}$$

Although equation 4.2 describes the calories in the food, the actual energetic profit that the bird obtains from its efforts is given by

$$E = F - C$$

where

E = energetic profit (in calories), and
C = the cost of obtaining food (in calories).

This latter value C is related to the fixed maintenance metabolism of the bird (which in this case is 10 cal) plus the searching and travelling costs; or

$$C = 10 + S + T$$

and therefore

$$E = 100P + 2SP + 0{\cdot}5STP + TP - 10P^2 - (10 + S + T) \qquad (4.3)$$

During the research to obtain equation 4.1, it was found that the bird did not expend more than 20 cal on searching and travelling costs. Thus there is a constraint such that

$$S + T = 20$$

If one wished to estimate the calorific value of a unit of prey that would maximise the energetic profit for the bird, it would be necessary to differentiate E in equation 4.3 with respect to P. Note, however, that this operation would involve differentiating E, which is a function of more than one variable. Thus, prior to estimating the extreme point for $\mathrm{d}E/\mathrm{d}P$, one must establish a technique for differentiating equations which contain more than one variable.

4.2. Partial Differentiation

If, in a function of several variables, all but one of the variables are held constant, then the function becomes a function of a single variable. Under this condition, the function can be differentiated in the normal way with respect to the single non-constant variable, and the resulting derivative is called the partial derivative of that variable.

Thus, if there is a function z which equals $\mathrm{f}(x, y)$, then one can obtain the partial derivative with respect to $\mathrm{f}(x)$ with y held constant, and the derivative with respect to $\mathrm{f}(y)$ with x held constant. These two partial derivatives are symbolised as f_x and f_y respectively, where

$$\mathrm{f}_x(x, y) = \lim_{\Delta x \to 0} \frac{\mathrm{f}(x + \Delta x, y) - \mathrm{f}(x, y)}{\Delta x} = \frac{\partial z}{\partial x}$$

and

$$f_y(x, y) = \underset{\Delta y \to 0}{\text{limit}} \frac{f(x, y + \Delta y) - f(x, y)}{\Delta y} = \frac{\partial z}{\partial y}$$

The symbol $\partial z/\partial x$ is also used to denote the partial differential of z with respect to x.

For example, if

$$z = f(x, y) = x^4 + 3y^2$$

then

$$f_x(x, y) = 4x^3$$

and

$$f_y(x, y) = 6y$$

Partial differentiation can be extended to as many variables as is desired, and the partial derivatives can be obtained by holding all but one of the variables constant. For example, if

$$z = f(x_1, x_2, x_3, y_1, y_2, y_3)$$
$$= x_1^2 x_2^4 + x_3^2 y_1^2 + y_2^6 y_3^5$$

then

$$f_{x_1} = 2x_1 x_2^4 \qquad f_{x_2} = 4x_1^2 x_2^3 \qquad f_{x_3} = 2x_3 y_1^2$$

$$f_{y_1} = 2x_3^2 y_1 \qquad f_{y_2} = 6y_2^5 y_3^5 \qquad f_{y_3} = 5y_2^6 y_3^4$$

Higher order derivatives can also be obtained by the same process. For example, if

$$z = f(x, y) = x^4 y^2$$

then

$$f_x = 4x^3 y^2 \text{ and } f_y = 2x^4 y$$

and

$$f_{xx} = 12x^2 y^2 \text{ and } f_{yy} = 2x^4$$

In addition,

$$f_{xy} = 8x^3 y \text{ and } f_{yx} = 8x^3 y$$

4.3. Extreme Points of Partial Derivatives

Given a function $z = f(x, y)$ that has a first-order derivative, then $f(x, y)$ has an extreme point (x_m, y_m) if

$$f_x(x_m, y_m) = 0 \text{ and } f_y(x_m, y_m) = 0$$

and if the condition

$$f_{xx}(x_m, y_m)\ f_{yy}(x_m, y_m) - f_{xy}^2(x_m, y_m) > 0$$

is fulfilled. Also, if both

$$f_{xx}(x_m, y_m) < 0 \text{ and } f_{yy}(x_m, y_m) < 0$$

the point is a maximum point; and if both

$$f_{xx}(x_m, y_m) > 0 \text{ and } f_{yy}(x_m, y_m) > 0$$

then it is a minimum point.

Example

If

$$z = f(x, y) = -2x^2 - 4x + 2xy - 3y^2 + 10y$$

then

$$f_x = -4x - 4 + 2y$$

and

$$f_y = 2x - 6y + 10$$

Setting f_x and f_y equal to zero, it is then possible to solve for the two unknowns x and y and this gives the result $x = 1$ and $y = 2$. To see if these values give an extreme point it is necessary to examine the second-order derivatives, namely

$$f_{xx} = -4 \qquad f_{yy} = -6 \qquad f_{xy} = 2$$

Now

$$f_{xx}\ f_{yy} - f_{xy}^2 = (-4)(-6) - (+2)^2 = 20$$

which is greater than zero, so the conditions for an extreme point are satisfied. In addition, $f_{xx} = -4$ and $f_{yy} = -6$ are both less than zero, and therefore the function $z = f(x, y)$ attains a local maximum at the point $(1, 2)$.

4.4. Solution of the Biological Example

The feeding activities of the bird (Section 4.1) were subject to the constraint that $S + T = 20$, and therefore

$$T = 20 - S \tag{4.4}$$

which can be substituted for T in equation 4.3 to give

$$E = 100P + 2SP + 0{\cdot}5SP(20 - S) + P(20 - S) - 10P^2 - 10 - S - (20 - S)$$

$$= 120P + 11SP - 0{\cdot}5S^2P - 10P^2 - 30 \tag{4.5}$$

Now E is the function of two variables (S and P), and to find the extreme value of E it is necessary to obtain the partial derivatives with respect to S and P, and to examine the value of these variables when their respective partial derivatives equal zero.

Thus, from equation 4.5,

$$E_P = 120 + 11S - 0{\cdot}5S^2 - 20P \tag{4.6}$$

and

$$E_S = 11P - SP \tag{4.7}$$

At the extreme point both these equations equal zero, and the values of S and P can be calculated. From equation 4.7, therefore,

$$11P - SP = 0$$

which can be rewritten as

$$P(11 - S) = 0$$

As it is unlikely that the prey will have a zero calorific value, then $11 - S$ must equal zero, or

$$S = 11$$

Similarly, from equation 4.6,

$$120 + 11S - 0{\cdot}5S^2 - 20P = 0$$

or

$$P = \frac{120 + 11S - 0{\cdot}5S^2}{20}$$

Substituting the value $S = 11$ gives

$$P = 9{\cdot}025$$

To find if E has an extreme point at the values of $S = 11$ and $P = 9{\cdot}025$, it is also necessary to consider the expression

$$E_{PP}\,E_{SS} - E_{PS}^2$$

which, from equations 4.6 and 4.7, is equal to

$$(-20)\,(-P) - (11 - S)^2$$

It has already been shown that $11 - S = 0$, and, when $P = 9{\cdot}025$, $20P$ is greater than zero. Therefore,

$$20P - (11 - S)^2 > 0$$

when $S = 11$ and $P = 9{\cdot}025$. Thus it can be concluded that there is an extreme value of E at the point $S = 11, P = 9{\cdot}025$. Further, because $E_{PP} = -20$ (which is less than zero), the point is a maximum.

Therefore, for the bird to maximise its energetic profit from the feeding activities under the constraint conditions, it must consume a prey which has a unit value of 9·025 cal.

4.5. Lagrange Multipliers

In the previous example it was possible to remove the variable T from equation 4.3 by the insertion of the constraint $T = 20 - S$. The reader will have probably noted that the insertion caused a certain degree of manipulation and, if the constraint had been any more complex, the manipulation of the equation would have become very cumbersome. To overcome this type of problem, it is possible to use an approach which involves Lagrange multipliers.

Consider the problem of finding an extreme point of the function $z = f(x, y)$ subject to the constraint $g(x, y) = 0$. The extreme point (x_m, y_m) must satisfy three equations such that $f_x(x_m, y_m) = 0$, $f_y(x_m, y_m) = 0$, and $g(x_m, y_m) = 0$. In such a situation, there are three equations to be satisfied by the two unknowns, and, when this occurs, the system is said to be overdetermined.

To find a solution to the system, an artificial unknown λ (the Lagrange multiplier) is introduced to produce the Lagrangian expression

$$f_\lambda = f(x, y) + \lambda g(x, y) \tag{4.8}$$

and, to satisfy the conditions for an extreme point, it is necessary that

$$\frac{\partial f_\lambda}{\partial x} = 0 \qquad \frac{\partial f_\lambda}{\partial y} = 0 \qquad \frac{\partial f_\lambda}{\partial \lambda} = 0$$

If these conditions are applied to the Lagrangian expression (equation 4.8), then

$$\frac{\partial f_\lambda}{\partial x} = f_x(x, y) + \lambda g_x(x, y) = 0 \tag{4.9}$$

$$\frac{\partial f_\lambda}{\partial y} = f_y(x, y) + \lambda g_y(x, y) = 0 \tag{4.10}$$

$$\frac{\partial f_\lambda}{\partial \lambda} = g(x, y) = 0 \tag{4.11}$$

For an extreme point to occur at $f(x_m, y_m)$, it is necessary to satisfy the equation

$$f_\lambda(x_m, y_m) = f(x_m, y_m) + \lambda g(x_m, y_m) \tag{4.12}$$

Example

Given the function

$$z = f(x, y) = 10xy$$

the aim is to find the extreme point subject to the constraint that

$$2x + 4y = 8$$

Now the Lagrangian expression is

$$f_\lambda = 10xy + \lambda(2x + 4y - 8)$$

and to satisfy the conditions for the extreme point:

$$\frac{\partial f_\lambda}{\partial x} = 10y + 2\lambda = 0 \left(\text{i.e. } y = \frac{-2\lambda}{10}\right) \tag{4.13}$$

$$\frac{\partial f_\lambda}{\partial y} = 10x + 4\lambda = 0 \left(\text{i.e. } x = \frac{-4\lambda}{10}\right) \tag{4.14}$$

$$\frac{\partial f_\lambda}{\partial \lambda} = 2x + 4y - 8 = 0 \tag{4.15}$$

By substituting the calculated values for x and y into the equation 4.15 to give

$$\frac{2(-4\lambda)}{10} + \frac{4(-2\lambda)}{10} - 8 = 0$$

it is possible to solve for λ, which yields

$$\lambda = -5$$

This value can be substituted into the expressions for x and y in equations 4.13 and 4.14 to give $x = 2$ and $y = 1$. Thus the extreme point of $10xy$ occurs at $x = 2, y = 1$, i.e. the extreme point has a value of 20.

4.6. An Alternative Approach to the Predator Strategy Problem

The relationship between energetic profit and the feeding variable was summarised in equation 4.3 as

$$E = 100P + 2SP + 0{\cdot}5STP + TP - 10P^2 - 10 - S - T$$

which was subject to the constraint $S + T = 20$.

The Lagrangian expression for this situation is

$$E_\lambda = E + \lambda(S + T - 20) \qquad (4.16)$$

and the conditions for the extreme point are

$$\frac{\partial E_\lambda}{\partial P} = 100 + 2S + 0{\cdot}5ST + T - 20P = 0 \qquad (4.17)$$

$$\frac{\partial E_\lambda}{\partial S} = 2P + 0{\cdot}5TP - 1 + \lambda = 0 \qquad (4.18)$$

$$\frac{\partial E_\lambda}{\partial T} = 0{\cdot}5SP + P - 1 + \lambda = 0 \qquad (4.19)$$

$$\frac{\partial E_\lambda}{\partial \lambda} = S + T - 20 = 0 \qquad (4.20)$$

Subtracting equation 4.19 from equation 4.18 gives

$$P - 0{\cdot}5SP + 0{\cdot}5TP = 0$$

or

$$P(1 - 0{\cdot}5S + 0{\cdot}5T) = 0$$

Since $P \neq 0$ (i.e. the prey must have a calorific value),

$$1 - 0{\cdot}5S + 0{\cdot}5T = 0$$

and therefore

$$0{\cdot}5S - 0{\cdot}5T = 1$$

or

$$T = S - 2$$

Substituting this value in equation 4.20 gives

$$S + (S - 2) - 20 = 0$$

or

$$S = 11$$

and, as $T = S - 2$, then

$$T = 9$$

Using these values in equation 4.17 yields

$$P = \frac{100 + 22 + 49{\cdot}5 + 9}{20}$$
$$= 9{\cdot}025$$

which is identical to the prey value calculated by the previous method as the unit value which would maximise the energetic profit of the bird. Note, however, that the answer obtained from the Lagrangian expression did not involve the substitution for T prior to the final analysis.

Exercises

1. Given the function

$$z = f(x_1, x_2, x_3, y_1, y_2, y_3) = x_1^2 x_2^3 + x_3^4 y_1^2 + y_2^3 y_3^2$$

find the values for $f_{x_1}, f_{x_2}, f_{x_3}, f_{y_1}, f_{y_2}$, and f_{y_3}.

2. Given the function

$$z = f(x, y) = x^2y^3$$

find the values for $f_x, f_{xx}, f_y, f_{yy}, f_{xy}$, and f_{yx}.

3. Given the function

$$z = f(x, y) = -x^2 - 2x + 2xy - 2y^2 + 10y$$

find the extreme point (x, y) where the function attains a local maximum.

4. Given the function

$$z = f(x, y) = 8xy$$

subject to the constraint $2x + y = 8$, use a Lagrangian expression to find the value of the extreme point for the function.

Five

INTEGRATION

5.1. Biological Example

A biologist studying the ecology of an estuarine mollusc took one metre square benthic samples every 3 m along a transect line which was 18 m in length. From the analysis of the samples, he found that the number of molluscs per square metre could be described by the equation

$$N = 18x - x^2 \tag{5.1}$$

where

N = number of molluscs per square metre, and
x = distance along the transect line (in metres).

By using equation 5.1 and plotting N against x (see Figure 5.1) it is apparent that the distribution curve of the molluscs is such that, as one moves from the beginning of the transect, the number of molluscs present increases towards a maximum at the point where $x = 9$. Past this point, the number of molluscs then decreases and eventually reaches zero at the point where $x = 18$.

Having analysed the samples and examined the above distribution curve, the biologist now wishes to estimate the total number of molluscs present in the area of benthic samples covered by his transect. This he can achieve by applying a basic technique of calculus, known as integration, to equation 5.1.

5.2. The Definite Integral

The area of a rectangle is equal to the base multiplied by the height, and this can be represented on a Cartesian co-ordinate system where height is a constant function of x [$y = f(x) = c$]. Both sides of a rectangle have the same height, and therefore, in Figure 5.2, $f(a) = f(b)$ and the area A is described by the equation

$$A = (b - a)\, f(a) = (b - a)\, f(b) \tag{5.2}$$

If $a = x_1$ and $b = x_1 + \Delta x$, then the area of the rectangle, from equation 5.2, is

$$A = (x_1 + \Delta x - x_1)\,f(a) = (x_1 + \Delta x - x_1)\,f(b)$$

which can be simplified to

$$A = \Delta x\ f(a) = \Delta x\ f(b)$$

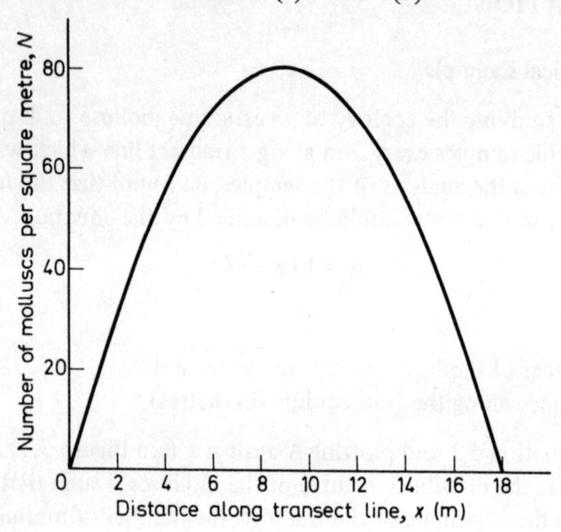

Figure 5.1

Now consider the function illustrated in Figure 5.3, where

$$\Delta x\ f(a) < \Delta x\ f(b)$$

because $f(a) < f(b)$. If the area under the curve is taken as equal to $\Delta x\ f(b)$, the error in this approximation is equivalent to the shaded portion in the diagram.

But consider what happens if the area under the curve between a and b in Figure 5.3 is divided into two rectangles R_1 and R_2 (Figure 5.4). If $f(z)$ is taken as the height of both rectangles, then the base of R_1 will equal $z - a$ and the base of R_2 will equal $b - z$. Thus the area under the curve can be approximated by the expression

$$(z - a)\,f(z) + (b - z)\,f(z)$$

and thus the shaded error portion has been reduced.

If one continues to subdivide the area under the curve into smaller and smaller rectangles, then the error continues to be reduced and the

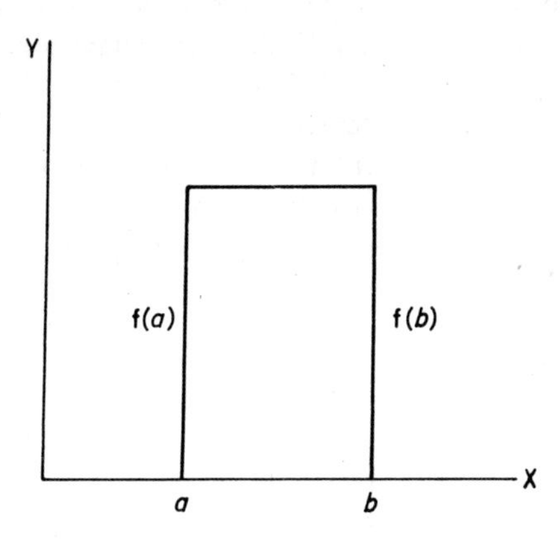

Figure 5.2

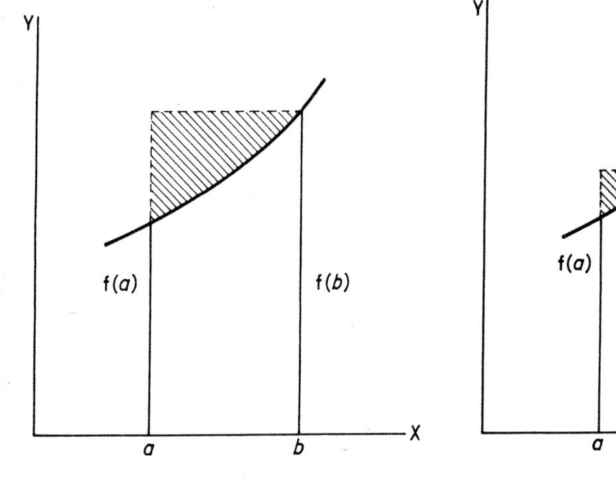

Figure 5.3

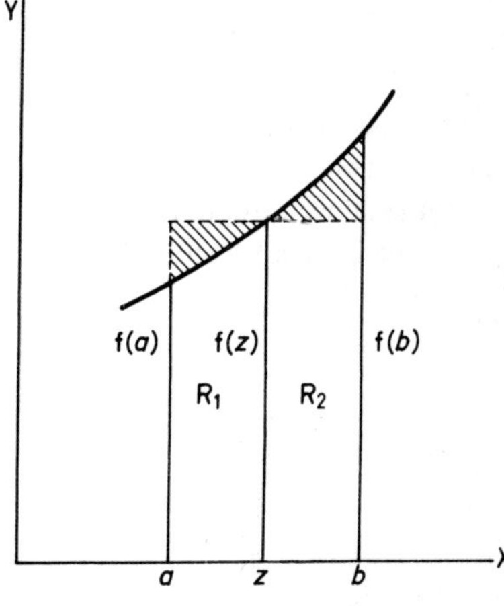

Figure 5.4

summation of the rectangles would begin to approach the true area under the curve.

This concept can be placed on a more formal mathematical basis by considering a function $y = f(x)$ which is continuous for $a \leqslant x \leqslant b$. The interval between a and b is divided into n subintervals by the $n - 1$ points $x_1, x_2, \ldots, x_{n-1}$, where $a = x_0$ and $b = x_n$. Now if an arbitrary point z_i is taken anywhere within the ith subinterval ($i = 1, 2, \ldots, n$), then the area of the ith strip can be approximated by that of the rectangle that has base $x_i - x_{i-1}$ and whose height is $f(z_i)$.

The sum of the areas of all such rectangles between $x = a$ and $x = b$ is therefore

$$(x_1 - x_0)\ f(z_1) + (x_2 - x_1)\ f(z_2) + (x_3 - x_2)\ f(z_3) + \ldots + (x_{n-1} - x_{n-2})\ f(z_{n-1}) + (x_n - x_{n-1})\ f(z_n)$$

which can be rewritten as

$$\sum_{i=1}^{n} (x_i - x_{i-1})\ f(z_i)$$

where the sigma notation denotes the summation of all such terms from $i = 1$ to $i = n$. Also, since

$$x_i - x_{i-1} = \Delta x$$

the summation expression can be written as

$$\sum_{i=1}^{n} \Delta x\ f(z_i)$$

As n (the number of subintervals) tends towards infinity (i.e. $n \to \infty$), the base of each subinterval will tend towards zero (i.e. $\Delta x \to 0$) and the above summation will approach the true area under the curve. This area is usually referred to as the definite integral and is written as

$$\int_a^b f(x)\,dx$$

Therefore

$$\int_a^b f(x)\,dx = \lim_{\substack{n \to \infty \\ \Delta x \to 0}} \sum_{i=1}^{n} \Delta x\ f(z_i)$$

Note that in the term

$$\int_a^b f(x)\,dx$$

f(x) is known as the integrand and x as the variable of integration, and the whole term denotes the limit of the area under the curve between $x = a$ and $x = b$.

Example

Considering the function $f(x) = x^2$ in the interval between $x = 0$ and $x = 1$, let $x_1 = 0{\cdot}1$ and $z_i = \frac{1}{2}(x_i + x_{i-1})$ for $i = 1, 2, \ldots, 10$. Then

$$x_0 = 0, x_1 = 0{\cdot}1, x_2 = 0{\cdot}2, \ldots, x_{10} = 1{\cdot}0$$

and

$$z_1 = 0{\cdot}05, z_2 = 0{\cdot}15, z_3 = 0{\cdot}25, \ldots, z_{10} = 0{\cdot}95$$

Thus the approximate area under the curve between $x = 0$ and $x = 1$ is

$$(x_1 - x_0)\, f(z_1) + (x_2 - x_1)\, f(z_2) + \ldots + (x_{10} - x_9)\, f(z_{10})$$

$$= (0{\cdot}1 - 0)\,(0{\cdot}05)^2 + (0{\cdot}2 - 0{\cdot}1)\,(0{\cdot}15)^2 + \ldots + (1{\cdot}0 - 0{\cdot}9)\,(0{\cdot}95)^2$$

$$= 0{\cdot}3325$$

Naturally this approximation would tend towards the true area under the curve as $\Delta x \to 0$, but the calculation would become rather lengthy unless one had a computer to do the operation. It is therefore necessary to find a method of directly estimating the value of an integral and so avoiding the summation.

5.3. Antiderivatives

Consider an expression

$$f_1(x) = 2x^3 + 3x^2$$

which is differentiable and has a derivative

$$f_1'(x) = 6x^2 + 6x$$

Now if an expression $z(x)$ exists which equals $6x^2 + 6x$, then $z(x) = f_1'(x)$ and $f_1(x)$ (i.e. $2x^3 + 3x^2$) is said to be an antiderivative of $z(x)$. Note, however, that the expression

$$f_2(x) = 2x^3 + 3x^2 + 6$$

is also an antiderivative of $z(x)$ because $f_2'(x) = 6x^2 + 6x$.

The only difference between $f_1(x)$ and $f_2(x)$ is that the latter expression has the additional term '+6', and this fact can be written as

$$f_2(x) = f_1(x) + C$$

where C is a constant, which in this case is 6.

Thus a general expression can be given for the antiderivative of the function $z(x)$, namely $f_1(x) + C$, where C is a constant.

5.4. The Fundamental Theorem of Integration

Consider an integral

$$\int_a^x f(u)\,du$$

which describes the area under a curve between the limits $u = a$ and $u = x$. Now if this integral is assumed to equal a function $F(x)$, then it is a property of integrals that the derivative of the function [i.e. $F'(x)$] is equal to the value of the integrand at its upper limits, i.e. $F'(x) = f(x)$.

For proof of this theorem, assume that

$$F(x) = \int_a^x f(u)\,du \tag{5.3}$$

Then

$$F'(x) = \lim_{\Delta x \to 0} \frac{F(x + \Delta x) - F(x)}{\Delta x}$$

or, from equation 5.3,

$$F'(x) = \lim_{\Delta x \to 0} \frac{\int_a^{x+\Delta x} F(u)\,du - \int_a^x F(u)\,du}{\Delta x}$$

$$= \lim_{\Delta x \to 0} \frac{\int_a^x F(u)\,du + \int_x^{x+\Delta x} F(u)\,du - \int_a^x F(u)\,du}{\Delta x}$$

$$= \lim_{\Delta x \to 0} \frac{1}{\Delta x} \int_x^{x+\Delta x} F(u)\,du \tag{5.4}$$

Now, from the initial discussion of integrals,

$$\int_{x}^{x+\Delta x} f(u)\, du$$

can be considered as the area of a rectangle of base Δx and height $f(x)$, and therefore equation 5.4 can be rewritten as

$$F'(x) = \underset{\Delta x \to 0}{\text{limit}} \frac{1}{\Delta x} [\Delta x \ f(x)]$$

$$= f(x)$$

Thus the derivative of the integral equals the value of the integrand at its upper limit.

However, remembering the comments upon the definition of antiderivatives, if $F'(x) = f(x)$, then the antiderivative of $f(x)$ will be $F(x) + C$, and therefore the integral

$$\int_{a}^{x} f(u)\, du = F(x) + C$$

From the above proof, it is now possible to derive the Newtonian fundamental theorem of calculus. Consider a continuous function $f(u)$ over an undefined interval. Let $u = a$ and $u = b$ be two points on that interval such that $a < b$, and let F be some function which is differentiable on this interval. Now from the above proof it is known that, for all values of x between a and b,

$$\int_{a}^{x} f(u)\, du = F(x) + C \tag{5.5}$$

and, if $x = a$, then

$$\int_{a}^{x} f(u)\, du = \int_{a}^{a} f(u)\, du = F(a) + C$$

Now it is an elementary property of integrals that

$$\int_{a}^{a} f(u)\, du = 0$$

and therefore

$$F(a) + C = 0$$

or

$$C = -F(a) \tag{5.6}$$

In addition,

$$\int_a^b f(u)\, du = F(b) + C$$

and thus, from equation 5.6,

$$\int_a^b f(u)\, du = F(b) - F(a) \tag{5.7}$$

which is the Newtonian theorem.

Having established this theorem, it is now possible to apply it to integration, i.e. the evaluation of definite integrals.

Consider the integral

$$\int_a^b f(x)\, dx = \int_a^b x^n\, dx \tag{5.8}$$

where the antiderivative of x^n is

$$\frac{x^{n+1}}{n+1} = F(x)$$

(The reader who is confused by this step may understand it by remembering that finding the antiderivative is the opposite process to differentiation.) Now, from equation 5.7,

$$\int_a^b f(x)\, dx = F(b) - F(a)$$

Therefore

$$\int_a^b x^n\, dx = \frac{b^{n+1}}{n+1} - \frac{a^{n+1}}{n+1} \tag{5.9}$$

Note that the right-hand side of equation 5.9 is often written as

$$\left.\frac{x^{n+1}}{n+1}\right|_a^b$$

where a and b are the lower and upper limits of the expression.

Example 1

Suppose that in equation 5.8, $a = 2$, $b = 4$, and $n = 2$; i.e.

$$\int_a^b f(x)dx = \int_2^4 x^2 dx$$

By integrating,

$$\int_a^b x^2 dx = \frac{b^3}{3} - \frac{a^3}{3}$$

$$= \frac{4^3}{3} - \frac{2^3}{3}$$

$$= 18\tfrac{2}{3}$$

Example 2

Consider the integral

$$\int_0^4 f(x)\,dx$$

where $f(x) = 6x^2 + 4x + 1$. In order to solve this expression, it is necessary to apply the property of integrals that, where $f(x) = g(x) + h(x)$, then

$$\int_0^4 f(x)dx = \int_0^4 g(x)dx + \int_0^4 h(x)dx$$

Thus

$$\int_0^4 (6x^2 + 4x + 1)\,dx = \int_0^4 6x^2\,dx + \int_0^4 4x\,dx + \int_0^4 1\,dx$$

The antiderivatives of these expressions are $2x^3$, $2x^2$, and x respectively. The last expression is more obvious if one remembers that, when $f(x) = x$, then $dy/dx = 1$. Thus integrating yields

$$2x^3\Big|_0^4 + 2x^2\Big|_0^4 + x\Big|_0^4 = (128 - 0) + (32 - 0) + (4 - 0)$$

$$= 164$$

The above examples have been concerned with definite integrals, i.e. integrals for which the limits have been defined. However, it is possible to have an integral where the limits are not defined, e.g.

$$\int f(x)\,dx$$

and this type of expression is known as an indefinite integral. It must be remembered in reference to the indefinite integral that the solution, because of the properties of antiderivatives, is always in the form $F(x) + C$, where C is the constant of integration.

For example, if

$$f(x) = x^2$$

then

$$\int f(x)\,dx = \int x^2\,dx$$
$$= \frac{1}{3}x^3 + C$$

Similarly, if

$$f(x) = \exp 2x$$

then

$$\int f(x)\,dx = \int (\exp 2x)\,dx$$
$$= \frac{1}{2}(\exp 2x) + C$$

If $f(x) = \sqrt[4]{x}$ [which can be written as $f(x) = x^{1/4}$], then

$$\int f(x)\,dx = \int x^{1/4}\,dx$$
$$= \frac{4}{5}x^{5/4} + C$$

5.5. Solution to the Mollusc Problem

The total number of molluscs N_t present in the area of the benthos covered by the transect (see Section 5.1) can be found by integrating the equation

$$N = 18x - x^2$$

and therefore

$$N_t = \int_0^{18} f(x)\,dx$$
$$= \int_0^{18} (18x - x^2)\,dx$$
$$= \int_0^{18} 18x\,dx - \int_0^{18} x^2\,dx$$

By integration,

$$N_t = 9x^2\Big|_0^{18} - \frac{x^3}{3}\Big|_0^{18}$$

$$= (9 \times 18^2 - 9 \times 0^2) - \left(\frac{18^3}{3} - \frac{0^3}{3}\right)$$

$$= 2916 - 1944$$

$$= 972$$

Therefore the total number of molluscs in the benthos is 972.

5.6. Other Methods of Solving Integrals

In previous sections, an integral could be solved because its antiderivative was easily recognised, and in many cases this approach is quite feasible. Consider, however, the integral

$$\int (x^3 + 2)^2 3x^2 \, dx$$

where the integrand $(x^3 + 2)^2 \, 3x^2$ is a complex expression. In this case one cannot guess the antiderivative, but it is possible instead to use the substitution method, as follows.

Let

$$u = x^3 + 2$$

Then

$$\frac{du}{dx} = 3x^2$$

or

$$du = 3x^2 \, dx$$

which is equivalent to the right-hand part of the required integral. Therefore the total expression can be rewritten as

$$\int u^2 \, du$$

which can be solved to give the result

$$\int u^2 \, du = \frac{1}{3} u^3 + C$$

Thus, replacing u by $x^3 + 2$,

$$\int (x^3 + 2)^2 \, 3x^2 \, dx = \frac{1}{3}(x^3 + 2)^3 + C$$

In the above example, the integration was accomplished by introducing a new variable u which simplified the original expression. This approach can often be applied, but consider the integral

$$\int \log_e x \, dx$$

In this case the substitution method is not applicable, and a technique known as the integration by parts has to be applied.

If f(x) and g(x) are two differentiable functions of x, then

$$\frac{d[f(x)\,g(x)]}{dx} = f(x)\frac{d[g(x)]}{dx} + g(x)\frac{d[f(x)]}{dx}$$

or

$$f(x)\frac{d[g(x)]}{dx} = \frac{d[f(x)\,g(x)]}{dx} - g(x)\frac{d[f(x)]}{dx}$$

and integration yields

$$\int f(x)\frac{d[g(x)]}{dx}\,dx = f(x)\,g(x) - \int g(x)\frac{d[f(x)]}{dx}\,dx \qquad (5.10)$$

Now if f(x) = u and g(x) = v, then equation 5.10 can be rewritten as

$$\int u \, dv = uv - \int v \, du \qquad (5.11)$$

This expression can be used to solve the integral

$$\int \log_e x \, dx$$

by putting $u = \log_e x$ and $v = x$. Then

$$\frac{du}{dx} = \frac{1}{x}$$

or

$$du = \frac{1}{x}\,dx$$

and $dv = dx$. Thus the required integral can be rewritten as

$$\int \log_e x \, dx = \int u \, dv$$

But, from equation 5.11,

$$\int u \, dv = uv - \int v \, du$$

Therefore

$$\int \log_e x \, dx = (\log_e x) \, x - \int x \left(\frac{1}{x}\right) dx$$

$$= (\log_e x) \, x - \int 1 \, dx$$

$$= x(\log_e x) - x + C$$

It is hoped that this chapter has introduced the reader to the basic techniques of integration. It should be pointed out, however, that if trouble is encountered in working on the solution of an integral, then possibly an answer may be found by using a published table of common integrals which can be found in most mathematical textbooks on calculus.

Exercises

Solve the following six indefinite integrals:

1. $\int 2x^2 \, dx$
2. $\int 5x^4 \, dx$
3. $\int x^{1/3} \, dx$
4. $\int (\exp x) \, dx$
5. $\int (\exp 3x) \, dx$
6. $\int (x^4 + 4)^2 \, 4x^3 \, dx$

7. Find the area under the curve defined by the integral

$$\int_0^3 (x^2 + 2x + 1) \, dx$$

Six

FIRST-ORDER DIFFERENTIAL EQUATIONS

6.1. Biological Example

If a species was placed in an unlimited environment, the increases in population size would probably follow a curve of the type shown in Figure 6.1, where the number of animals N is plotted against time t. If it was known that the number of animals at any point was some function of time, then to find the number after a given time, one could simply integrate the expression

$$\int f(t)\, dt$$

However, the rate of increase of animals in an unlimited environment is described by the classic biological equation

$$\frac{dN}{dt} = rN \tag{6.1}$$

where

N = number of animals,
t = time, and
r = innate capacity of increase.

At first inspection it would appear that one might use the derivative rN as the integrand and solve for the number of animals from the expression

$$\int rN\, dt$$

Unfortunately, however, although r is a constant and can be taken outside of the integral, N is a dependent variable and therefore direct integration of the expression is not possible.

In fact, any equation which expresses a relationship between unknowns of one or several variables, and which contains both the functions and their derivatives, is known as a differential equation. When there is only one independent variable, the equation is an ordinary differential equation: equation 6.1 is of this type because it expresses a relationship between the dependent variable N (the number of animals), the independent variable t (time), and the derivative rN.

Thus, if one wishes to calculate the number of animals in the environment after a certain time, it is necessary to obtain a solution or

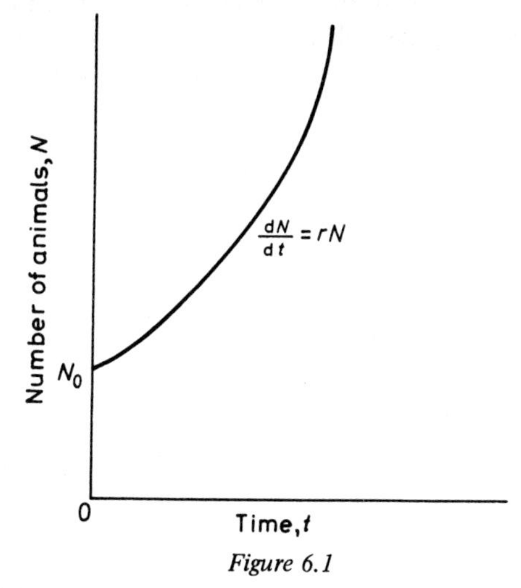

Figure 6.1

'integral' of the differential equation. The actual solving of differential equations is not the simplest of tasks because there are numerous different approaches and the texts on the subject are usually written with the assumption that the reader has a thorough understanding of basic calculus.

To illustrate the type of approach used in solving differential equations, the remaining sections in this chapter contain details of two techniques that can be adopted. It is suggested, however, that if the reader can expect to be involved in work that will require solving many differential equations, then he should consider embarking upon a formal course in the subject or endeavour to obtain the loyal services of a mathematician.

6.2. The Order of Differential Equations

A differential equation is said to be of the order n if the highest derivative in the expression is the nth derivative. For example, the equation

$$y' + xy + x = 0$$

is a first-order differential equation because the highest derivative is y'.

Normally, a first-order differential equation is written in the form

$$F(x, y, y') = 0$$

where F is a given function of the three variables, x being the independent variable, y the unknown dependent variable, and y' the derivative dy/dx. Thus, in attempting to solve the equation, one solves for $y = g(x)$ such that

$$F(x, y, y') = F[x, g(x), g'(x)] = 0$$

i.e. one expresses y in terms of x.

6.3. The Separation of Variables Technique

Given

$$\frac{dy}{y} = 3x^2\, dx \tag{6.2}$$

the left-hand side of the equation involves y and the right-hand side x, and therefore the variables are separate. Since

$$\frac{dy}{y} \equiv \frac{1}{y}\, dy$$

integration of equation 6.2 yields

$$\log_e y = x^3 + c \tag{6.3}$$

where c is the constant of integration, and, as it is an arbitrary value, it could be set equal to $\log_e a$. Then equation 6.3 becomes

$$\log_e y = x^3 + \log_e a$$

or

$$y = a \exp x^3 \tag{6.4}$$

However, now consider the equation

$$\frac{dy}{dx} = 3x^2 y \tag{6.5}$$

In this case the variables are not separate, but they can be separated by multiplying throughout by $\mathrm{d}x/y$ to give

$$\frac{\mathrm{d}y}{y} = 3x^2 \, \mathrm{d}x \qquad (6.6)$$

Thus equation 6.6 is now in the same form as equation 6.2. The integration step can now be carried out to yield

$$y = a \exp x^3$$

Thus the differential equation 6.5 has been solved by expressing y in terms of x.

6.4. The Solution of the Animal Numbers Problem

The relationship between the number of animals N and time t is given as (equation 6.1)

$$\frac{\mathrm{d}N}{\mathrm{d}t} = rN$$

Multiplying through by $\mathrm{d}t/N$ yields

$$\frac{\mathrm{d}N}{N} = r \, \mathrm{d}t \qquad (6.7)$$

Having separated the variables, it is possible to integrate equation 6.7, remembering that r is a constant and that the integral on the right-hand side is therefore of the form

$$r \int 1 \, \mathrm{d}t$$

Thus

$$\log_e N = rt + c$$

and, setting the arbitrary constant equal to $\log_e N_0$ (where N_0 is the original number of animals introduced into the environment), then

$$N = N_0 \exp rt$$

where N is the number of animals present at any time t.

6.5. Linear Differential Equations with a Constant Coefficient

Given an equation

$$Y' + Y\ \mathrm{P}(x) = \mathrm{Q}(x) \tag{6.8}$$

where $\mathrm{P}(x)$ and $\mathrm{Q}(x)$ are functions of x only, then the equation is said to be linear. As the highest derivative in the equation is the first-order derivative Y', equation 6.8 is said to be a first-order linear differential equation.

If, in equation 6.8, $\mathrm{P}(x)$ is a constant such that $\mathrm{P}(x) = \alpha$, then the equation is a first-order linear differential equation with a constant coefficient and may be rewritten as

$$Y' + Y\alpha = \mathrm{Q}(x) \tag{6.9}$$

where $\mathrm{Q}(x)$ is a continuous function of x. If $\mathrm{Q}(x)$ is set equal to zero, equation 6.9 yields the homogeneous form

$$y' + y\alpha = 0 \tag{6.10}$$

which can be solved for y and y' in terms of x.

First, multiplying equation 6.10 through by $\mathrm{d}x/y$ (remembering that $y' = \mathrm{d}y/\mathrm{d}x$) gives

$$\frac{\mathrm{d}y}{y} + \alpha\ \mathrm{d}x = 0$$

or

$$\frac{\mathrm{d}y}{y} = -\alpha\ \mathrm{d}x$$

Then integration yields

$$\log_e y = -\alpha x + C$$

or, if the arbitrary constant C is taken as $\log_e c$,

$$y = c \exp(-\alpha x) \tag{6.11}$$

Equation 6.11 is said to be the general solution of the homogeneous equation.

Now assume that $\overline{Y}(x)$ is some function that satisfies equation 6.9 for all values of x ($\overline{Y}$ is said to be a particular solution). Then

$$Y = c \exp(-\alpha x) + \overline{Y} \tag{6.12}$$

because the general solution of a differential equation is equal to the sum of the general solution of the homogeneous equation and a particular solution. To demonstrate this, differentiate equation 6.12 with respect to x:

$$Y' = -\alpha c \exp(-\alpha x) + \bar{Y}' \tag{6.13}$$

Substituting equation 6.12 and 6.13 into equation 6.9 gives

$$[-\alpha c \exp(-\alpha x) + \bar{Y}'] + \alpha[c \exp(-\alpha x) + \bar{Y}] = Q(x)$$

and thus

$$-\alpha c \exp(-\alpha x) + \bar{Y}' + \alpha c \exp(-\alpha x) + \alpha\bar{Y} = Q(x)$$

or

$$\bar{Y}' + \bar{Y}\alpha = Q(x) \tag{6.14}$$

Thus it is proved that, if one takes a general solution of an homogeneous equation and adds a particular solution, then one obtains a general solution for the original equation.

To evaluate the arbitrary constant c of integration it is necessary to set $Y(x) = Y_0$ for $x = 0$ and, as

$$Y(x) = c \exp(-\alpha x) + \bar{Y}(x) \tag{6.15}$$

then

$$\begin{aligned} Y_0 &= c \exp(-\alpha 0) + \bar{Y}(0) \\ &= c + \bar{Y}(0) \end{aligned}$$

and thus

$$c = Y_0 - \bar{Y}(0) \tag{6.16}$$

Substituting into equation 6.15 gives

$$Y = [Y_0 - \bar{Y}(0)] \exp(-\alpha x) + \bar{Y} \tag{6.17}$$

This apparently circuitous proof can probably be better understood by applying it to an example.

6.6. Biological Example

There are 50 animals in a defined area at the beginning of a study and these are dying at a rate of 20% of the total population per week. In addition, more of the species are entering the area and the immigration rate is 20 animals per week. The biologist doing this study now wishes to use the data to estimate the population size in the area after 2 weeks.

The information can be summarised by the equation

$$\frac{dN}{dt} = I(t) - D(t)\,N \tag{6.18}$$

where

dN/dt = rate of increase of population,
$I(t)$ = level of immigration, and
$D(t)\,N$ = number dying.

Now it is known that the death rate is a constant proportion of the animals present and therefore $D(t)$ can be set equal to the constant α. Also, it is known that 20 new animals are entering the area per week and thus $I(t) = 20$. Equation 6.18 can therefore be rewritten as

$$\frac{dN}{dt} = 20 - \alpha N$$

or

$$\frac{dN}{dt} + \alpha N = 20$$

or

$$N' + \alpha N = 20 \tag{6.19}$$

which is a first-order linear differential equation with a constant coefficient. Thus, if $I(t)$ is set equal to zero, the homogeneous form of equation 6.19 is

$$n' + \alpha n = 0$$

and, from the previous section, it is known that

$$n = c \exp(-\alpha t) \tag{6.20}$$

which is the general solution of the homogeneous equation.

Now if a particular solution of equation 6.19 for all values of t is

$$\bar{N}(t) = at$$

then

$$\frac{d\bar{N}}{dt} = a$$

Substitution into equation 6.19 yields

$$a + \alpha a t = 20$$

or

$$a = \frac{20}{1 + \alpha t} \tag{6.21}$$

Thus, from equation 6.21, the particular solution $\bar{N}(t)$ can be rewritten as

$$\bar{N}(t) = \frac{20t}{1 + \alpha t}$$

As the general solution of an equation is equal to the sum of the general solution of the homogeneous equation and a particular solution,

$$N(t) = c \exp(-\alpha t) + \bar{N}(t)$$

$$= c \exp(-\alpha t) + \frac{20t}{1 + \alpha t} \tag{6.22}$$

To find the value of the arbitrary constant c it is necessary to examine equation 6.22 at the initial conditions (where $t = 0$), at which point there are 50 animals in the area. Thus

$$N_0 = 50 = c \exp(\alpha 0) + \frac{20 \times 0}{1 + \alpha 0}$$

or

$$c = 50$$

and inserting this into equation 6.22 yields

$$N(t) = 50 \exp(-\alpha t) + \frac{20t}{1 + \alpha t} \tag{6.23}$$

Equation 6.23 can be used to estimate the number of animals in the area after two weeks. At this time $t = 2$, and $\alpha = 0{\cdot}2$ as 20% of the animals are dying per week. Thus

$$N(t) = 50 \exp(-0{\cdot}2 \times 2) + \frac{20 \times 2}{1 + (0{\cdot}2 \times 2)}$$

$$= 50 \times 0{\cdot}67 + \frac{40}{1{\cdot}4}$$

and therefore it can be concluded that there will be approximately 62 animals in the area at the end of two weeks.

Seven

DIFFERENCE EQUATIONS

7.1. Biological Example

Given that

N_t = number of animals present at time t, and
N_{t-1} = number present in the previous generation,

then the growth rate G of the population is described by the equation

$$G = \frac{N_t - N_{t-1}}{N_{t-1}} \tag{7.1}$$

Thus, if N_{t-1} has a value of 50 and N_t a value of 60, then the growth rate as a percentage is

$$G = \frac{60 - 50}{50} \times 100\%$$

$$= 20\%$$

If this problem is now approached on the basis of knowing the growth rate, then

$$20 = \frac{N_t - N_{t-1}}{N_{t-1}} \times 100$$

or

$$20N_{t-1} = 100N_t - 100N_{t-1}$$

Thus

$$100N_t = 120N_{t-1}$$

which can be rewritten as

$$N_t = aN_{t-1} \tag{7.2}$$

where a is a constant, and in this example

$$a = \frac{120}{100} = 1{\cdot}2$$

The expression $N_t = aN_{t-1}$ is a difference equation because it relates the dependent variable N in terms of the time period. In the above example the time period is $t-1$, and therefore equation 7.2 is known as a first-order difference equation. If, for example, the equation had had the form

$$N_t = aN_{t-1} + bN_{t-2}$$

then it would be a second-order equation because it involved both $t-1$ and $t-2$.

Before the arrival of computers, it was very difficult to solve difference equations because the approach often involves iterative curve fitting. With the advent of the tireless digital computer, however, such iteration is now practical, and ecologists are beginning to use difference equations in problem solving. Thus the reader should be aware of the existence of such equations, and one approach to their solution is discussed in the next section.

7.2. First-Order Linear Difference Equations

A general form of the first-order linear difference equation with constant coefficients is

$$Y_t = aY_{t-1} + \mathrm{f}(t) \tag{7.3}$$

This is said to have constant coefficients because all the coefficients of the dependent variable Y are constants and do not involve t.

To solve equation 7.3, it is possible to use the same approach that was applied to differential equations, i.e. a general solution may be found which is equal to the sum of the general solution of the homogeneous equation and a particular solution. Thus, setting $\mathrm{f}(t)$ equal to zero, the homogeneous form of equation 7.3 is

$$y_t = ay_{t-1} \tag{7.4}$$

which has the general solution

$$y_t = Ca^t \tag{7.5}$$

To show that the general solution of equation 7.4 is of this type, equation 7.5 can be rewritten as

$$\begin{aligned} y_t &= Ca^t \\ &= a(Ca^{t-1}) \\ &= ay_{t-1} \end{aligned}$$

Thus equation 7.5 is the general solution to the homogeneous equation 7.4.

If $\overline{Y}_t$ is taken as a particular solution of equation 7.3, then the general solution of this equation is equal to the sum of the general solution of the homogeneous equation and the particular solution. Thus

$$Y_t = Ca^t + \overline{Y}_t \tag{7.6}$$

To evaluate C, it is necessary to consider equation 7.6 at the initial conditions where $t = 0$. Then

$$Y_0 = Ca^0 + \overline{Y}_0$$

where $\overline{Y}_0$ is the value of $\overline{Y}_t$ when $t = 0$. Thus

$$C = Y_0 - \overline{Y}_0$$

and equation 7.6 can be rewritten as

$$Y_t = (Y_0 - \overline{Y}_0)a^t + \overline{Y}_t \tag{7.7}$$

7.3. Biological Example

The study of an insect population in a limited environment revealed that, as the numbers increased, the amount of food per insect decreased, and from this evidence it was concluded that the number dying was related to the number present (Figure 7.1). This relationship could be expressed by the equation

$$N_D = A + aN_t \tag{7.8}$$

where

N_D = number dying,
N_t = number present in generation t, and
A and a are constants.

The insects, the adults of which are not capable of overwintering, reproduce by laying eggs which hatch the next year. In addition, the study revealed that, if there were a large number of adults present in any one year, then the number of eggs that were laid decreased because more energy was spent on intraspecific competition (Figure 7.2). As the eggs hatch during the next year, the initial size of the population in any year is a function of the number of adults present in the previous generation, i.e.

$$N_B = B + bN_{t-1} \tag{7.9}$$

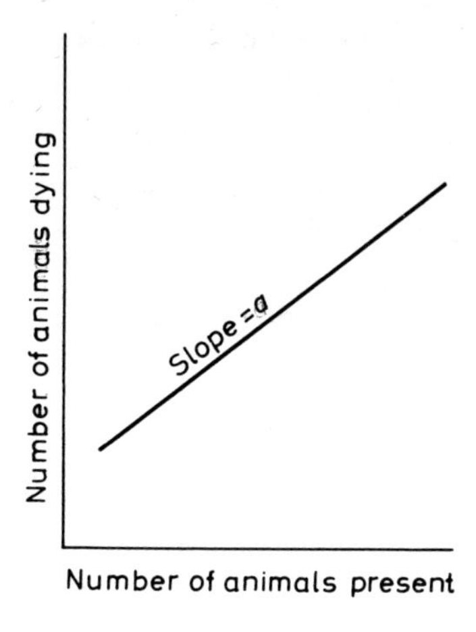

Figure 7.1

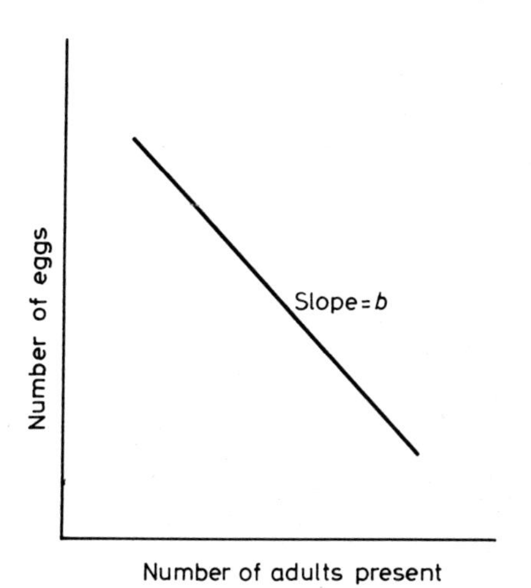

Figure 7.2

where

N_{B} = number of emerging young,
N_{t-1} = number of adults present in generation $t-1$, and
B and b are constants.

From this information it is possible to examine the number of animals present and to predict the number of animals present in future generations. Accepting that the numbers of births and deaths are equal,

$$N_{\mathrm{D}}(N_t) = N_{\mathrm{B}}(N_{t-1})$$

or, from equations 7.8 and 7.9,

$$A + aN_t = B + bN_{t-1}$$

Therefore

$$N_t = \frac{b}{a}N_{t-1} + \frac{B-A}{a} \tag{7.10}$$

which is a first-order linear difference equation with constant coefficients of the dependent variable N.

Thus, if

$$\frac{B-A}{a} = 0$$

then the homogeneous form of equation 7.10 is

$$n_t = \frac{b}{a}n_{t-1} \tag{7.11}$$

which has the general solution

$$n_t = C\left(\frac{b}{a}\right)^t \tag{7.12}$$

where C is an arbitrary constant.

Now assume that $\overline{N}$ is a particular solution of equation 7.10 for all values of t. Then

$$\overline{N} = \frac{b}{a}\overline{N} + \frac{B-A}{a}$$

or

$$\overline{N}\left(1 - \frac{b}{a}\right) = \frac{B-A}{a}$$

Therefore

$$\bar{N}\left(\frac{a-b}{a}\right) = \frac{B-A}{a}$$

and thus

$$\bar{N} = \frac{B-A}{a-b}$$

A general solution to equation 7.10 can now be obtained from the sum of the general solution of the homogeneous equation and the particular solution, i.e.

$$N_t = C\left(\frac{b}{a}\right)^t + \bar{N}$$

$$= C\left(\frac{b}{a}\right)^t + \frac{B-A}{a-b} \qquad (7.13)$$

To calculate C, it is necessary to set t equal to zero, at which point the number of animals in the population is N_0. Thus

$$C = N_0 - \bar{N}$$

$$= N_0 - \frac{B-A}{a-b}$$

which can be substituted into equation 7.13 to give

$$N_t = \left(N_0 - \frac{B-A}{a-b}\right)\left(\frac{b}{a}\right)^t + \frac{B-A}{a-b}$$

or

$$N_t = (N_0 - \bar{N})\left(\frac{b}{a}\right)^t + \bar{N} \qquad (7.14)$$

for $t = 0, 1, 2, \ldots, n$.

This final equation can now be used to predict the number of animals in the population at any time t. Note that, in evaluating N_t,

1. The birth rate has a negative slope b because the number of young decreases with an increasing N_t.
2. The death rate has a positive slope a because the number of deaths increases with an increasing N_t.

In evaluating N_t it will be assumed that the particular solution $\bar{N}$ represents the optimum size for the population in the area.

There are three possible conditions to consider in evaluating N_t.

Condition 1

Assume that

$$|b| > |a|$$

where $| \ |$ denotes the absolute value of the term enclosed. Then, as $t \to \infty$, the value of $|(b/a)^t|$ will increase indefinitely. However, as b is negative and a is positive, the sign of $(b/a)^t$ will be alternately negative and positive for $t = 1, 2, \ldots, n$.

Now, if the population size at the beginning of the study is smaller than the optimum size N, then

$$N_0 - \bar{N} < 0 \text{ (i.e. negative)}$$

At the end of the first year after study commences, i.e. where $t = 1$,

$$\left(\frac{b}{a}\right)^t < 0 \text{ (i.e. negative)}$$

But, from equation 7.14,

$$N_t - \bar{N} = (N_0 - \bar{N})\left(\frac{b}{a}\right)^t \tag{7.15}$$

Therefore

$$N_1 - \bar{N} > 0 \text{ (i.e. positive)}$$

and, as $|b| > |a|$,

$$|N_1 - \bar{N}| > |N_0 - \bar{N}|$$

After the second year ($t = 2$),

$$\left(\frac{b}{a}\right)^t > 0 \text{ (i.e. positive)}$$

and $N_0 - \bar{N}$ is negative as before. Therefore, from equation 7.15,

$$N_2 - \bar{N} < 0 \text{ (i.e. negative)}$$

although

$$|N_2 - \bar{N}| > |N_1 - \bar{N}|$$

At the end of each year, the absolute value $|N_t - \overline{N}|$ will become further and further away from $\overline{N}$ and, as $N_t - \overline{N}$ is alternately positive

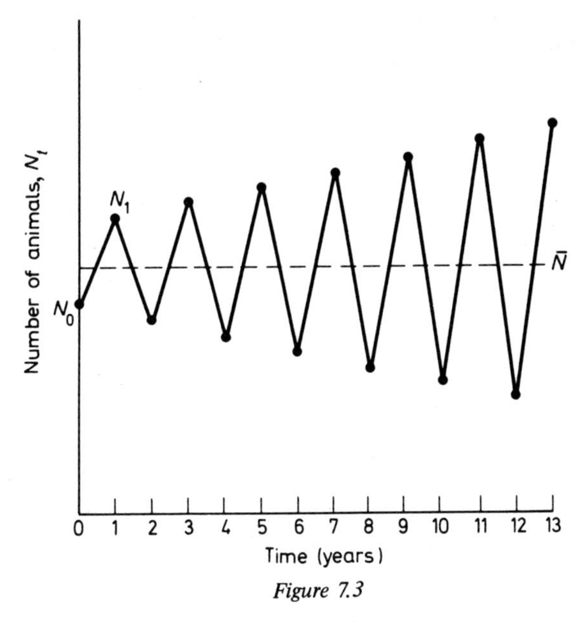

Figure 7.3

and negative, the system will oscillate in the form illustrated in Figure 7.3.

Condition 2

Assume that

$$|b| = |a|$$

Then, as $t \to \infty$, $(b/a)^t$ will be equal to -1 and $+1$ alternately.

Now, if the population size at the beginning of the study is smaller than the optimum size $\overline{N}$, then

$$N_0 - \overline{N} < 0$$

and, at the end of the first year after the study commences ($t = 1$),

$$\left(\frac{b}{a}\right)^t = -1$$

Therefore, from equation 7.15,

$$N_1 - \bar{N} = -(N_0 - \bar{N})$$

Thus

$$N_1 - \bar{N} > 0$$

and

$$|N_1 - \bar{N}| = |N_0 - \bar{N}|$$

At the end of the second year ($t = 2$),

$$\left(\frac{b}{a}\right)^t = +1$$

and, as $N_0 - \bar{N}$ is negative, then from equation 7.15,

$$N_2 - \bar{N} < 0$$

and

$$N_2 - \bar{N} = N_0 - \bar{N}$$

At the end of each year, the absolute value of $N_t - \bar{N}$ will be constant and equal to $|N_0 - \bar{N}|$, and therefore the system will oscillate regularly around $\bar{N}$ (Figure 7.4).

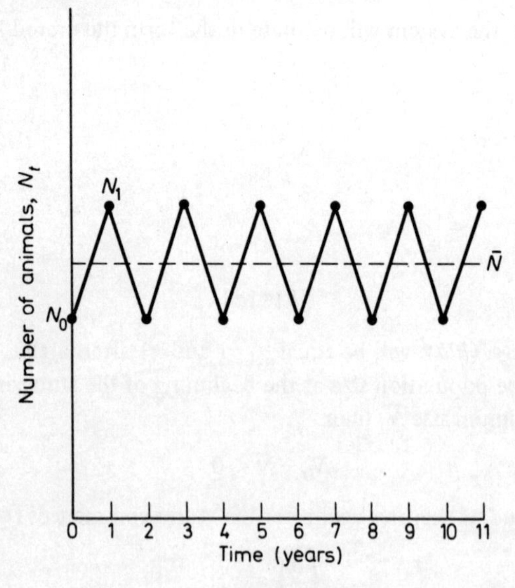

Figure 7.4

Condition 3

Assume that

$$|b| < |a|$$

Then, as $t \to \infty$, $|(b/a)^t|$ will approach zero.

Now, if the population size at the beginning of the study is smaller than $\bar{N}$, then

$$N_0 - \bar{N} < 0$$

and, at the end of the first year after the study commences,

$$\left(\frac{b}{a}\right)^t < 0$$

Therefore, from equation 7.15,

$$N_1 - \bar{N} > 0$$

and, as $|b/a| < 1$,

$$|N_1 - \bar{N}| < |N_0 - \bar{N}|$$

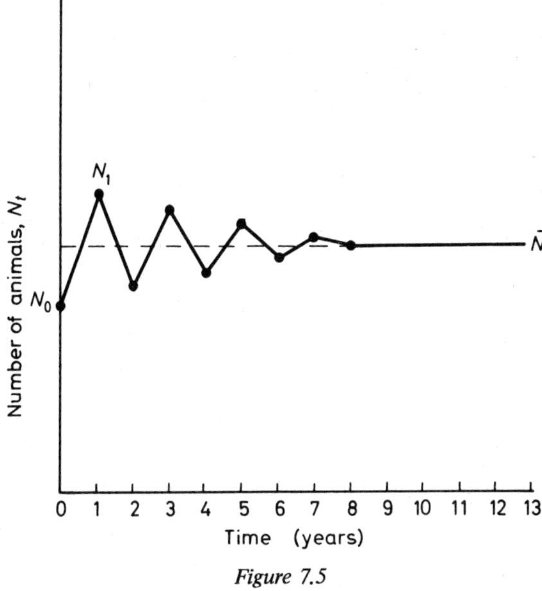

Figure 7.5

At the end of the second year ($t = 2$),

$$\left|\left(\frac{b}{a}\right)^t\right| < \left|\frac{b}{a}\right|$$

and, as $N_0 - \overline{N}$ is negative, then from equation 7.15,

$$N_2 - \overline{N} < 0$$

and

$$|N_2 - \overline{N}| < |N_0 - \overline{N}|$$

At the end of each year, the absolute value $|N_t - \overline{N}|$ will be smaller and smaller until the system approaches the optimum value $\overline{N}$ (Figure 7.5).

Consequently it is apparent from the above analysis that, if there is to be a constant number of animals in the population from year to year, then it is necessary that the birth rate should decrease at a slower rate b with increasing N_t than the rate a of increase of the death rate with increasing N_t, i.e. the situation demonstrated by Condition 3.

Some readers may have found difficulty following the discussion of the three populations described above. The author suggests that these readers may understand the mechanics of equation 7.15 if they actually fit values to the terms in the equation. For example, set N_0 at 80 and $\overline{N}$ at 100, and take small values such as 4 and 2 to fit the appropriate conditions set for a and b.

Eight

MATRICES

8.1. Biological Example

A study of an insect population gave the data shown in Table 8.1 for the total number of each life stage observed in four areas during a year. Sampling a year later revealed that the numbers of each stage had risen to those shown in Table 8.2. These sets of data were used to calculate: (a) the total number of each stage observed during the two years, and (b) the change in numbers of each stage over the two-year period.

Although this is a simple problem, the analysis can be made easier by the use of matrices.

Table 8.1

	Study area			
	1	*2*	*3*	*4*
Eggs	2221	5172	1006	1556
1st instar	1103	2561	511	1234
2nd instar	617	1338	326	818
3rd instar	528	1124	278	646
4th instar	430	986	187	550
5th instar	312	727	114	414
Adults	202	542	106	309

Table 8.2

	Study area			
	1	*2*	*3*	*4*
Eggs	3674	6352	1206	1781
1st instar	1812	3106	564	1319
2nd instar	819	1518	352	864
3rd instar	545	1220	299	682
4th instar	462	1006	201	591
5th instar	329	798	166	454
Adults	236	604	122	339

8.2. The Definition of a Matrix

A matrix (plural – matrices) is a rectangular array of numbers arranged in rows and columns. For example,

$$\mathbf{A} = \begin{bmatrix} 4 & 2 \\ 3 & 6 \end{bmatrix}$$

is a matrix, and the integers in the matrix are called the elements of the matrix. Note that the matrix is bounded by rectangular brackets and that an upper case letter is used to denote the array.

The dimensions of a matrix determine its order which is designated as 'n by m', where n is the number of rows (which is always stated first) and m is the number of columns. For example,

$$\mathbf{A} = \begin{bmatrix} 2 & 1 & 1 \\ 2 & 1 & 3 \end{bmatrix}$$

is a 2 by 3 matrix because there are 2 rows (2 1 1; 2 1 3) and 3 columns (2 2; 1 1; 1 3).

When the number of rows equals the number of columns (i.e. $n = m$), then the matrix is called a square matrix. If the matrix consists of a single column, then it is called a column matrix (or column vector), and, if the matrix consists of a single row, it is a row matrix (or row vector). For instance,

$$\mathbf{A} = \begin{bmatrix} 4 & 2 \\ 3 & 6 \end{bmatrix}$$

is a square matrix ($n = m = 2$);

$$\mathbf{A} = \begin{bmatrix} 1 \\ 2 \\ 3 \end{bmatrix}$$

is a column matrix ($n = 3,\ m = 1$); and

$$\mathbf{A} = [1 \quad 2 \quad 3]$$

is a row matrix ($n = 1,\ m = 3$).

The position of any element in a matrix may be defined by its row and column notation. For example, element a_{34} would be found in the third row of the fourth column, and any element in a matrix is usually denoted by a_{ij}, where i is the row and j is the column. This notation can probably be clarified in an example:

$$\mathbf{A} = \begin{bmatrix} 4 & 3 & 5 & 7 & 9 & 5 \\ 2 & 2 & 4 & 3 & 1 & 1 \\ 1 & 8 & 7 & 6 & 4 & 9 \end{bmatrix}$$

can be rewritten as

$$\mathbf{A} = \begin{bmatrix} a_{11} & a_{12} & a_{13} & a_{14} & a_{15} & a_{16} \\ a_{21} & a_{22} & a_{23} & a_{24} & a_{25} & a_{26} \\ a_{31} & a_{32} & a_{33} & a_{34} & a_{35} & a_{36} \end{bmatrix}$$

or

$$\mathbf{A} = [a_{ij}]_{(n,\, m)}$$

where

$$\begin{aligned} i &= 1, 2, \ldots, n \\ j &= 1, 2, \ldots, m \\ n &= 3 \\ m &= 6 \end{aligned}$$

8.3. Addition and Subtraction of Matrices

Matrix addition can be performed only when the two matrices to be added are of the same order, i.e. where there are the same number of rows and columns in the two matrices. Given

$$\mathbf{A} = [a_{ij}]_{(n,\, m)}$$

and

$$\mathbf{B} = [b_{ij}]_{(n,\, m)}$$

then

$$\mathbf{A} + \mathbf{B} = [a_{ij} + b_{ij}]_{(n,\, m)}$$

For example,

$$\begin{bmatrix} 2 & 3 \\ 1 & 4 \end{bmatrix} + \begin{bmatrix} 6 & 7 \\ 8 & 9 \end{bmatrix} = \begin{bmatrix} (2+6) & (3+7) \\ (1+8) & (4+9) \end{bmatrix}$$

$$= \begin{bmatrix} 8 & 10 \\ 9 & 13 \end{bmatrix}$$

But

$$\begin{bmatrix} 4 & 3 \\ 3 & 2 \end{bmatrix} + \begin{bmatrix} 5 & 6 & 7 \\ 6 & 7 & 8 \end{bmatrix}$$

has no solution because the two matrices are not conformable for addition.

Matrix subtraction can also be performed only when the two matrices are of the same order. Thus, given

$$\mathbf{A} = [a_{ij}]_{(n,\, m)}$$

and

$$\mathbf{B} = [b_{ij}]_{(n,\, m)}$$

then

$$\begin{aligned} \mathbf{A} - \mathbf{B} &= \mathbf{A} + (-\mathbf{B}) \\ &= [a_{ij} - b_{ij}]_{(n,\, m)} \end{aligned}$$

For example,

$$\begin{aligned} \begin{bmatrix} 6 & 9 \\ 5 & 8 \end{bmatrix} - \begin{bmatrix} 3 & 4 \\ 3 & 2 \end{bmatrix} &= \begin{bmatrix} (6-3) & (9-4) \\ (5-3) & (8-2) \end{bmatrix} \\ &= \begin{bmatrix} 3 & 5 \\ 2 & 6 \end{bmatrix} \end{aligned}$$

8.4. Solution of Biological Example

The numbers of each life stage in each area can be considered as two matrices such that

$$\mathbf{A} = \begin{bmatrix} 2221 & 5172 & 1006 & 1556 \\ 1103 & 2561 & 511 & 1234 \\ 617 & 1338 & 326 & 818 \\ 528 & 1124 & 278 & 646 \\ 430 & 986 & 187 & 550 \\ 312 & 727 & 114 & 414 \\ 202 & 542 & 106 & 309 \end{bmatrix}$$

and

$$\mathbf{B} = \begin{bmatrix} 3674 & 6352 & 1206 & 1781 \\ 1812 & 3106 & 564 & 1319 \\ 819 & 1518 & 352 & 864 \\ 545 & 1220 & 299 & 682 \\ 462 & 1006 & 201 & 591 \\ 329 & 798 & 166 & 454 \\ 236 & 604 & 122 & 339 \end{bmatrix}$$

Then the total number of each stage observed over the two-year period can be found from the summation of $\mathbf{A}$ and $\mathbf{B}$. Thus $\mathbf{A} + \mathbf{B}$ is equal to

$$\begin{bmatrix} (2221 + 3674) & (5172 + 6352) & (1006 + 1206) & (1556 + 1781) \\ (1103 + 1812) & (2561 + 3106) & (511 + 564) & (1234 + 1319) \\ (617 + 819) & (1338 + 1518) & (326 + 352) & (818 + 864) \\ (528 + 545) & (1124 + 1220) & (278 + 299) & (646 + 682) \\ (430 + 462) & (986 + 1006) & (187 + 201) & (550 + 591) \\ (312 + 329) & (727 + 798) & (114 + 166) & (414 + 454) \\ (202 + 236) & (542 + 604) & (106 + 122) & (309 + 339) \end{bmatrix}$$

which gives the result

$$\begin{bmatrix} 5895 & 11524 & 2212 & 3337 \\ 2915 & 5667 & 1075 & 2553 \\ 1436 & 2856 & 678 & 1682 \\ 1073 & 2344 & 577 & 1328 \\ 892 & 1992 & 388 & 1141 \\ 641 & 1525 & 280 & 868 \\ 438 & 1146 & 228 & 648 \end{bmatrix} \begin{matrix} \text{Eggs} \\ \text{1st instar} \\ \text{2nd instar} \\ \text{3rd instar} \\ \text{4th instar} \\ \text{5th instar} \\ \text{Adults} \end{matrix}$$

In addition, the increase in population size for each life stage in each area can be found from $\mathbf{B} - \mathbf{A}$. Therefore the increase is

$$\mathbf{B} - \mathbf{A} = \mathbf{C}$$

where

$$\mathbf{C} = \begin{bmatrix} 1453 & 1180 & 200 & 225 \\ 709 & 545 & 53 & 85 \\ 202 & 180 & 26 & 46 \\ 17 & 96 & 21 & 36 \\ 32 & 20 & 14 & 41 \\ 17 & 71 & 52 & 40 \\ 34 & 62 & 16 & 30 \end{bmatrix}$$

During the study, data had also been collected from which it was possible to calculate the average weight of one individual of each life stage. The research worker on the project then wished to use this information to calculate the increase in biomass in each area during the second year as a result of the rise in numbers of each life stage.

To answer this problem, it is necessary to understand the process of multiplying matrices.

8.5. The Multiplication of Matrices

If $\mathbf{A}$ is an n by m matrix and $\mathbf{B}$ is another n by m matrix, then the product $\mathbf{AB} = \mathbf{C}$, where every entry c_{ij} of $\mathbf{C}$ is obtained by multiplying the entries of the ith row of $\mathbf{A}$ by the corresponding entries of the jth column of $\mathbf{B}$ and summing the product. Note, however, that for multiplication to be possible, the matrices must be conformable, i.e. the number of columns of $\mathbf{A}$ must equal the number of rows in $\mathbf{B}$.

Consider, for example, the two matrices

$$\mathbf{A} = \begin{bmatrix} 6 & 2 & 1 \\ 3 & 1 & 2 \\ 2 & 4 & 1 \end{bmatrix} \quad \mathbf{B} = \begin{bmatrix} 3 & 6 \\ 2 & 1 \\ 1 & 2 \end{bmatrix}$$

Now $\mathbf{A}$ is a 3 by 3 matrix and $\mathbf{B}$ is a 3 by 2 matrix, and therefore the two matrices are conformable for multiplication because the number of columns in $\mathbf{A}$ is 3 and the number of rows in $\mathbf{B}$ is also 3. Thus

$$\mathbf{AB} = \begin{bmatrix} (6 \times 3 + 2 \times 2 + 1 \times 1) & (6 \times 6 + 2 \times 1 + 1 \times 2) \\ (3 \times 3 + 1 \times 2 + 2 \times 1) & (3 \times 6 + 1 \times 1 + 2 \times 2) \\ (2 \times 3 + 4 \times 2 + 1 \times 1) & (2 \times 6 + 4 \times 1 + 1 \times 2) \end{bmatrix}$$

$$= \begin{bmatrix} 23 & 40 \\ 13 & 23 \\ 15 & 18 \end{bmatrix}$$

But consider now the two matrices

$$\mathbf{A} = \begin{bmatrix} 2 & 1 & 2 \\ 2 & 2 & 1 \\ 1 & 0 & 6 \end{bmatrix} \quad \mathbf{B} = \begin{bmatrix} 6 & 4 \\ 1 & 2 \end{bmatrix}$$

As $\mathbf{A}$ is a 3 by 3 matrix and $\mathbf{B}$ is a 2 by 2 matrix, the multiplication is not possible because the number of columns in $\mathbf{A}$ (i.e. 3) is greater than the number of rows in $\mathbf{B}$ (i.e. 2).

When matrices are to be multiplied, one often encounters the Σ (sigma) notation, which is merely an abbreviation to signify that values are to be summed. For example,

$$1 + 2 + 3 + 4 + 5 \equiv \sum_{x=1}^{5} x$$

(see also Section 5.2). The notation can be applied to matrices, as in the example

$$a_{11}b_{11} + a_{12}b_{21} + a_{13}b_{31} = \sum_{x=1}^{3} a_{1x}b_{x1}$$

which is the entry in the first row and column of a matrix which is the product of the multiplication of a matrix **A** and a matrix **B**.

Thus the product of two 2 by 2 matrices

$$\mathbf{A\,B} = \begin{bmatrix} a_{11} & a_{12} \\ a_{21} & a_{22} \end{bmatrix} \begin{bmatrix} b_{11} & b_{12} \\ b_{21} & b_{22} \end{bmatrix}$$

$$= \begin{bmatrix} (a_{11}b_{11} + a_{12}b_{21}) & (a_{11}b_{12} + a_{12}b_{22}) \\ (a_{21}b_{11} + a_{22}b_{21}) & (a_{21}b_{12} + a_{22}b_{22}) \end{bmatrix}$$

can be written in sigma notation as

$$\mathbf{A\,B} = \begin{bmatrix} \sum_{x=1}^{2} a_{1x}b_{x1} & \sum_{x=1}^{2} a_{1x}b_{x2} \\ \sum_{x=1}^{2} a_{2x}b_{x1} & \sum_{x=1}^{2} a_{2x}b_{x2} \end{bmatrix}$$

$$= \begin{bmatrix} \sum_{x=1}^{2} a_{ix}b_{xj} \end{bmatrix} \quad (i = 1, 2;\ j = 1, 2)$$

8.6. Solution to the Biomass Increase Problem

The data on the average weight of one individual of each life stage are as given in Table 8.3, which for the purpose of the multiplication can be written as a 1 by 7 matrix.

Table 8.3

	Egg	*1st instar*	*2nd instar*	*3rd instar*	*4th instar*	*5th instar*	*Adult*
Weight (g)	0·01	0·04	0·09	0·24	0·64	1·60	2·20

In order to multiply the weights by the life stages, it is necessary to transpose matrix **C** of Section 8.4 such that

$$\mathbf{C}^{\mathrm{T}} = \begin{bmatrix} 1453 & 709 & 202 & 17 & 32 & 17 & 34 \\ 1180 & 545 & 180 & 96 & 20 & 71 & 62 \\ 200 & 53 & 26 & 21 & 14 & 52 & 16 \\ 225 & 85 & 46 & 36 & 41 & 40 & 30 \end{bmatrix}$$

($\mathbf{C}^{\mathrm{T}}$ is the symbol used to signify that the matrix is a transpose of matrix **C**.)

Thus the increase in biomass in each area can be calculated by the multiplication of

$$\begin{bmatrix} 1453 & 709 & 202 & 17 & 32 & 17 & 34 \\ 1180 & 545 & 180 & 96 & 20 & 71 & 62 \\ 200 & 53 & 26 & 21 & 14 & 52 & 16 \\ 225 & 85 & 46 & 36 & 41 & 40 & 30 \end{bmatrix} \begin{bmatrix} 0{\cdot}01 \\ 0{\cdot}04 \\ 0{\cdot}09 \\ 0{\cdot}24 \\ 0{\cdot}64 \\ 1{\cdot}60 \\ 2{\cdot}20 \end{bmatrix}$$

Note that the multiplication is possible because the number of columns in the first matrix equals the number of rows in the second matrix. The result of the multiplication is

$$\begin{bmatrix} 14{\cdot}53 + 28{\cdot}36 + 18{\cdot}18 + 4{\cdot}08 + 20{\cdot}48 + 27{\cdot}20 + 74{\cdot}80 \\ 11{\cdot}80 + 21{\cdot}80 + 16{\cdot}20 + 23{\cdot}04 + 12{\cdot}80 + 113{\cdot}60 + 136{\cdot}40 \\ 2{\cdot}00 + 2{\cdot}12 + 2{\cdot}34 + 5{\cdot}04 + 8{\cdot}96 + 83{\cdot}20 + 35{\cdot}20 \\ 2{\cdot}25 + 3{\cdot}40 + 4{\cdot}14 + 8{\cdot}64 + 26{\cdot}24 + 64{\cdot}00 + 66{\cdot}00 \end{bmatrix} = \begin{bmatrix} 187{\cdot}63 \\ 335{\cdot}64 \\ 138{\cdot}88 \\ 174{\cdot}67 \end{bmatrix}$$

Thus the total increase in biomass in the four areas in the second year is

Area 1	Area 2	Area 3	Area 4
187·63	335·64	138·88	174·67

Having completed this chapter, the reader may consider that the above examples could be solved without recourse to matrices. The advantages of the matrix approach probably become more apparent when one is faced with a wealth of data, because then the analysis can be made much easier by setting it up in a matrix form. In addition, the reader should be aware of matrix notation because it is often used when tabulated data are analysed on a digital computer.

Accepting that the above examples may not require analysis by matrices, nevertheless, when one is presented with n linear equations, the use of matrices greatly aids their simultaneous solution. This type of problem is dealt with in the next chapter.

Exercises

1. Calculate the sum

$$\begin{bmatrix} 3 & 2 & 6 \\ 4 & 7 & 2 \end{bmatrix} + \begin{bmatrix} 2 & 1 & 2 \\ 1 & 1 & 4 \end{bmatrix}$$

2. Calculate the result of the subtraction

$$\begin{bmatrix} 4 & 7 \\ 9 & 8 \end{bmatrix} - \begin{bmatrix} 2 & 1 \\ 3 & 7 \end{bmatrix}$$

3. Multiply

$$\begin{bmatrix} 4 & 3 \\ 2 & 2 \end{bmatrix} \text{by} \begin{bmatrix} 6 & 1 \\ 3 & 2 \end{bmatrix}$$

4. Multiply

$$\begin{bmatrix} a_{11} & a_{12} & a_{13} \\ a_{21} & a_{22} & a_{23} \end{bmatrix} \text{ by } \begin{bmatrix} b_{11} & b_{12} \\ b_{21} & b_{22} \\ b_{31} & b_{32} \end{bmatrix}$$

and rewrite the product in sigma notation.

Nine

SIMULTANEOUS LINEAR EQUATIONS

9.1. Biological Example

In feeding experiments carried out on four trout, the fish were permitted to feed and then the stomach contents were extracted from the fish before digestion had commenced. After ingestion, the food had been broken into pieces during its passage to the stomach, and therefore it was not feasible to subdivide the diets to find the number of calories of each of the species consumed. It was possible, however, to calculate the number of each food species consumed by identifying and counting the head capsules which remained intact. After this count had been made, the total stomach contents of each fish was 'bombed' and the calorific value obtained. The resulting data were as given in Table 9.1.

Table 9.1

	Number of species consumed				*Total calorific value of stomach contents*
	Species 1	*Species 2*	*Species 3*	*Species 4*	
Fish 1	18	12	8	6	1110
Fish 2	14	8	10	16	1340
Fish 3	8	10	14	14	1320
Fish 4	2	2	16	16	1130

Now, if x_1, x_2, x_3, and x_4 denote the unknown calorific values of one individual of each of the prey species, then the data can be rewritten in the form of four linear equations such that

$$\begin{aligned}
18x_1 + 12x_2 + 8x_3 + 6x_4 &= 1110\\
14x_1 + 8x_2 + 10x_3 + 16x_4 &= 1340\\
8x_1 + 10x_2 + 14x_3 + 14x_4 &= 1320\\
2x_1 + 2x_2 + 16x_3 + 16x_4 &= 1130
\end{aligned}$$

It is possible to solve for the unknown calorific values of the prey species by applying matrix algebra techniques.

9.2. Determinants and the Application of Cramer's Rule

If an operation is carried out whereby all the elements within a square matrix **A** are described by a single value (or scalar), then this calculated value is known as the determinant of the matrix and is usually written as det A or $|A|$. Also, given the matrix

$$\mathbf{A} = \begin{bmatrix} a_{11} & a_{12} & a_{13} \\ a_{21} & a_{22} & a_{23} \\ a_{31} & a_{32} & a_{33} \end{bmatrix}$$

the determinant A can be written in the form

$$A = \begin{vmatrix} a_{11} & a_{12} & a_{13} \\ a_{21} & a_{22} & a_{23} \\ a_{31} & a_{32} & a_{33} \end{vmatrix}$$

To establish a general procedure for evaluating determinants, it is first necessary to define the minor and the cofactor of an element of a square matrix.

A minor of an element a_{ij} of a square matrix is the determinant of the submatrix which is obtained by deleting the ith row and the jth column. Thus, given the matrix

$$\mathbf{A} = \begin{bmatrix} a_{11} & a_{12} & a_{13} \\ a_{21} & a_{22} & a_{23} \\ a_{31} & a_{32} & a_{33} \end{bmatrix}$$

then the minor of the element a_{31} is

$$\begin{vmatrix} a_{12} & a_{13} \\ a_{22} & a_{23} \end{vmatrix}$$

or, in other words, the minor of a_{31} is obtained by deleting the third row and the first column.

The cofactor of an element a_{ij} of a square matrix is the product of the minor and $(-1)^{i+j}$, and is denoted by A_{ij}. Thus, given the minor of a_{31} as above, then the cofactor of a_{31} is

$$A_{31} = (-1)^{3+1} \begin{vmatrix} a_{12} & a_{13} \\ a_{22} & a_{23} \end{vmatrix} \qquad (9.1)$$

Now the value of a 2 by 2 minor is equal to the product of the top left-hand corner and the bottom right-hand corner, minus the product

product of the top right-hand corner and the bottom left-hand corner. Thus the value of the minor of a_{31} is

$$a_{12}\,a_{23} - a_{13}\,a_{22}$$

Therefore equation 9.1 can be rewritten as

$$A_{31} = (-1)^{3+1}\,(a_{12}\,a_{23} - a_{13}\,a_{22})$$

It is now possible to evaluate any determinant because the value of a determinant is equal to the sum of all the products obtained by multiplying each element in a column (or row) by its cofactor. Thus the value of the above third-order determinant is

$$\det A = a_{11}\,(-1)^{1+1}\,(a_{22}\,a_{33} - a_{23}\,a_{32}) + a_{21}\,(-1)^{2+1}\,(a_{12}\,a_{33} - a_{13}\,a_{32}) + a_{31}\,(-1)^{3+1}\,(a_{12}\,a_{23} - a_{13}\,a_{22})$$

In this case, the cofactor expansion was carried out on column 1, but the same value for the determinant would have been obtained if the expansion had been using column 2 or 3 or any one of the rows.

In general, the value of the nth-order determinant $A = |a_{ij}|_n$ is the sum of all terms of the form $(-1)^x\,a_{nj_n}$, where the two subscripts n and j_n assume all possible arrangements such that each column is represented exactly once in each term of the sum, and where x is the number of interchanges necessary to bring the subscripts into a natural order.

Example 1

Given the determinant

$$A = \begin{vmatrix} 1 & 2 & 3 \\ 1 & 3 & 4 \\ 1 & 2 & 2 \end{vmatrix}$$

expanding about column 1 yields

$$(1)\,(-1)^{1+1}\,(3 \times 2 - 4 \times 2) + (1)\,(-1)^{2+1}\,(2 \times 2 - 3 \times 2) + (1)\,(-1)^{3+1}\,(2 \times 4 - 3 \times 3))$$

$$= (+1)\,(6 - 8) + (-1)\,(4 - 6) + (+1)\,(8 - 9)$$

$$= -2 + 2 - 1$$

$$= -1$$

Example 2

To evaluate the determinant

$$A = \begin{vmatrix} 1 & 2 & 3 & 4 \\ 1 & 3 & 4 & 0 \\ 0 & 2 & 2 & 0 \\ 0 & 1 & 2 & 0 \end{vmatrix}$$

first note that expansion about the fourth column will greatly reduce the operation because three of the elements are zero.

Thus

$$\det A = (4)(-1)^{1+4} \begin{vmatrix} 1 & 3 & 4 \\ 0 & 2 & 2 \\ 0 & 1 & 2 \end{vmatrix} + (0)(-1)^{2+4} \begin{vmatrix} 1 & 2 & 3 \\ 0 & 2 & 2 \\ 0 & 1 & 2 \end{vmatrix}$$

$$+ (0)(-1)^{3+4} \begin{vmatrix} 1 & 2 & 3 \\ 1 & 3 & 4 \\ 0 & 1 & 2 \end{vmatrix} + (0)(-1)^{4+4} \begin{vmatrix} 1 & 2 & 3 \\ 1 & 3 & 4 \\ 0 & 2 & 2 \end{vmatrix}$$

As all but the cofactor A_{41} are equal to zero,

$$\det A = (-4) \begin{vmatrix} 1 & 3 & 4 \\ 0 & 2 & 2 \\ 0 & 1 & 2 \end{vmatrix}$$

and expanding the third-order sub-determinant about column 1 gives

$$\begin{aligned} \det A &= (-4)\,[(1)(-1)^{1+1}(2 \times 2 - 2 \times 1) + 0 + 0] \\ &= (-4)(+1)(+2) \\ &= -8 \end{aligned}$$

Example 3

Consider now the determinant

$$A = \begin{vmatrix} 1 & 2 & 3 & 6 \\ 1 & 5 & 2 & 4 \\ 2 & 6 & 4 & 8 \\ 1 & 7 & 1 & 3 \end{vmatrix}$$

where the expansion around any column or row would be a long process. To simplify the operation, one can use the property that, if a determinant B is obtained from a determinant A by adding, to each element of a column or row in A, a constant multiple of the corresponding element of another row or column, then

$$\det B = \det A$$

Thus in the example above, if the third column is multiplied by −2 and the product added to the fourth column, then

$$\begin{vmatrix} 1 & 2 & 3 & 6 \\ 1 & 5 & 2 & 4 \\ 2 & 6 & 4 & 8 \\ 1 & 7 & 1 & 3 \end{vmatrix}_{(-2C_3 + C_4)} = \begin{vmatrix} 1 & 2 & 3 & 0 \\ 1 & 5 & 2 & 0 \\ 2 & 6 & 4 & 0 \\ 1 & 7 & 1 & 1 \end{vmatrix}$$

Expanding about a_{44} gives

$$\det A = (+1) \begin{vmatrix} 1 & 2 & 3 \\ 1 & 5 & 2 \\ 2 & 6 & 4 \end{vmatrix}$$

which, by the operation $(-2C_1 + C_2)$, becomes

$$\det A = (+1) \begin{vmatrix} 1 & 0 & 3 \\ 1 & 3 & 2 \\ 2 & 2 & 4 \end{vmatrix}$$

Finally, by expanding about the second column,

$$\begin{aligned} \det A &= (+1)\,[(+3)\,(1 \times 4 - 3 \times 2) + (-2)\,(1 \times 2 - 1 \times 3)] \\ &= (+3)\,(-2) + (-2)\,(-1) \\ &= -4 \end{aligned}$$

A technique for evaluating determinants having been established, it is now possible to solve linear equations by applying Cramer's rule. Given, for example, the equations

$$\begin{aligned} x_1 + x_2 + x_3 &= 7 \\ x_1 + 2x_2 + x_3 &= 9 \\ x_1 + x_2 + 2x_3 &= 11 \end{aligned} \tag{9.2}$$

then this array can be rewritten as

$$A\mathbf{X} = \mathbf{B} \tag{9.3}$$

where

$$A = \begin{vmatrix} 1 & 1 & 1 \\ 1 & 2 & 1 \\ 1 & 1 & 2 \end{vmatrix} \quad \mathbf{X} = \begin{bmatrix} x_1 \\ x_2 \\ x_3 \end{bmatrix} \quad \mathbf{B} = \begin{bmatrix} 7 \\ 9 \\ 11 \end{bmatrix} \tag{9.4}$$

If ^jA is used to denote a determinant obtained from A by replacing column j of A by matrix $\mathbf{B}$, then by Cramer's rule, if $\det A \neq 0$ (in which case the determinant is said to be non-singular), the system $A\mathbf{X} = \mathbf{B}$ has exactly one solution such that

$$x_j = \frac{\det {}^jA}{\det A}$$

where $j = 1, 2, 3$.

Thus, to apply the rule to equations 9.2, it is first necessary to calculate $\det A$ to find if it is non-singular. Evaluation of A gives

$$\det A = 1$$

i.e. $\det A \neq 0$, so the expression $A\mathbf{X} = \mathbf{B}$ must have exactly one solution. Thus

$$x_1 = \frac{\begin{vmatrix} 7 & 1 & 1 \\ 9 & 2 & 1 \\ 11 & 1 & 2 \end{vmatrix}}{\det A}$$

$$= \frac{7(4-1) - 9(2-1) + 11(1-2)}{1}$$

$$= 1$$

Note that, to solve for x_1, column 1 of determinant A has been replaced by matrix $\mathbf{B}$. Similarly,

$$x_2 = \frac{\begin{vmatrix} 1 & 7 & 1 \\ 1 & 9 & 1 \\ 1 & 1 & 2 \end{vmatrix}}{\det A}$$

$$= \frac{(18-1) - (14-1) + (7-9)}{1}$$

$$= 2$$

and

$$x_3 = \frac{\begin{vmatrix} 1 & 1 & 7 \\ 1 & 2 & 9 \\ 1 & 1 & 11 \end{vmatrix}}{\det A}$$

$$= \frac{(22-9) - (11-7) + (9-14)}{1}$$

$$= 4$$

Thus the solution of simultaneous equations 9.2 is $x_1 = 1, x_2 = 2, x_3 = 4$.

9.3. Solution of Linear Equations by the Use of the Inverse

Another approach to solving linear equations is the use of the inverse, and to calculate this it is first necessary to define the cofactor and adjoint matrices of any matrix. This approach demands that the matrix must be square and non-singular.

The cofactor of a matrix, which should not be confused with the cofactor of any element of that matrix, is the array represented by the cofactors of all the elements a_{ij} in that matrix. Thus the cofactor of the matrix

$$\mathbf{A} = \begin{bmatrix} 1 & 1 & 1 \\ 1 & 2 & 1 \\ 1 & 1 & 2 \end{bmatrix} \tag{9.5}$$

is given by

$$\text{cof } \mathbf{A} = \begin{bmatrix} (+1)\begin{vmatrix} 2 & 1 \\ 1 & 2 \end{vmatrix} & (-1)\begin{vmatrix} 1 & 1 \\ 1 & 2 \end{vmatrix} & (+1)\begin{vmatrix} 1 & 2 \\ 1 & 1 \end{vmatrix} \\ (-1)\begin{vmatrix} 1 & 1 \\ 1 & 2 \end{vmatrix} & (+1)\begin{vmatrix} 1 & 1 \\ 1 & 2 \end{vmatrix} & (-1)\begin{vmatrix} 1 & 1 \\ 1 & 1 \end{vmatrix} \\ (+1)\begin{vmatrix} 1 & 1 \\ 2 & 1 \end{vmatrix} & (-1)\begin{vmatrix} 1 & 1 \\ 1 & 1 \end{vmatrix} & (+1)\begin{vmatrix} 1 & 1 \\ 1 & 2 \end{vmatrix} \end{bmatrix}$$

$$= \begin{bmatrix} 3 & -1 & -1 \\ -1 & 1 & 0 \\ -1 & 0 & 1 \end{bmatrix}$$

The adjoint of a matrix is the transpose of the cofactor of that matrix. Thus

$$\text{adj } \mathbf{A} = (\text{cof } \mathbf{A})^{\mathrm{T}} = \begin{bmatrix} 3 & -1 & -1 \\ -1 & 1 & 0 \\ -1 & 0 & 1 \end{bmatrix}$$

In this example, the transpose has the same array as the cofactor, but this is not always the case (see, for example, the transposition of $\mathbf{C}$ in Section 8.6).

Once the adjoint of a matrix has been obtained, the inverse can be calculated because the inverse of $\mathbf{A}$ (notation $\mathbf{A}^{-1}$) is given by

$$\mathbf{A}^{-1} = \frac{\text{adj } \mathbf{A}}{\det A}$$

Thus the inverse of the matrix **A** given in equation 9.5 is

$$\mathbf{A}^{-1} = \frac{\begin{bmatrix} 3 & -1 & -1 \\ -1 & 1 & 0 \\ -1 & 0 & 1 \end{bmatrix}}{\begin{vmatrix} 1 & 1 & 1 \\ 1 & 2 & 1 \\ 1 & 1 & 2 \end{vmatrix}} = \begin{bmatrix} 3 & -1 & -1 \\ -1 & 1 & 0 \\ -1 & 0 & 1 \end{bmatrix}$$

The inverse can be used to solve the linear equations because, in a system $A\mathbf{X} = \mathbf{B}$ (see equation 9.3) where A is non-singular,

$$\mathbf{X} = A^{-1}\mathbf{B}$$

(A^{-1} denotes the determinant of the inverse matrix $\mathbf{A}^{-1}$). Thus, from equations 9.4,

$$\mathbf{X} = \begin{bmatrix} x_1 \\ x_2 \\ x_3 \end{bmatrix} = \begin{vmatrix} 3 & -1 & -1 \\ -1 & 1 & 0 \\ -1 & 0 & 1 \end{vmatrix} \begin{bmatrix} 7 \\ 9 \\ 11 \end{bmatrix}$$

Therefore, multiplying the determinant A^{-1} by the matrix $\mathbf{B}$,

$$\begin{bmatrix} x_1 \\ x_2 \\ x_3 \end{bmatrix} = \begin{bmatrix} 21 - 9 - 11 \\ -7 + 9 \;\; -0 \\ -7 + 0 + 11 \end{bmatrix} = \begin{bmatrix} 1 \\ 2 \\ 4 \end{bmatrix}$$

which gives the result $x_1 = 1, x_2 = 2$, and $x_3 = 4$.

9.4. Solution of the Biological Example

The simultaneous equations in Section 9.1 can be rewritten in the form

$$A\mathbf{X} = \mathbf{B}$$

where

$$A = \begin{vmatrix} 18 & 12 & 8 & 6 \\ 14 & 8 & 10 & 16 \\ 8 & 10 & 14 & 14 \\ 2 & 2 & 16 & 16 \end{vmatrix} \qquad \mathbf{X} = \begin{bmatrix} x_1 \\ x_2 \\ x_3 \\ x_4 \end{bmatrix} \qquad \mathbf{B} = \begin{bmatrix} 1110 \\ 1340 \\ 1320 \\ 1130 \end{bmatrix}$$

The equations can then be solved by the application of Cramer's rule or by using the inverse $\mathbf{A}^{-1}$.

First, however, it is necessary to establish that A is non-singular: this condition is met because

$$\det A = 8880$$

9.4.1. Application of Cramer's Rule

By applying Cramer's rule,

$$x_1 = \frac{\begin{vmatrix} 1110 & 12 & 8 & 6 \\ 1340 & 8 & 10 & 16 \\ 1320 & 10 & 14 & 14 \\ 1130 & 2 & 16 & 16 \end{vmatrix}}{8880} = \frac{177\,600}{8880} = 20$$

Similarly,

$$x_2 = \frac{\begin{vmatrix} 18 & 1110 & 8 & 6 \\ 14 & 1340 & 10 & 16 \\ 8 & 1320 & 14 & 14 \\ 2 & 1130 & 16 & 16 \end{vmatrix}}{8880} = \frac{222\,000}{8880} = 25$$

Also

$$x_3 = \frac{\begin{vmatrix} 18 & 12 & 1110 & 6 \\ 14 & 8 & 1340 & 16 \\ 8 & 10 & 1320 & 14 \\ 2 & 2 & 1130 & 16 \end{vmatrix}}{8880} = \frac{266\,400}{8880} = 30$$

and

$$x_4 = \frac{\begin{vmatrix} 18 & 12 & 8 & 1110 \\ 14 & 8 & 10 & 1340 \\ 8 & 10 & 14 & 1320 \\ 2 & 2 & 16 & 1130 \end{vmatrix}}{8880} = \frac{310\,800}{8880} = 35$$

Thus the calorific values of individuals of the four species are 20 cal, 25 cal, 30 cal, and 35 cal.

9.4.2. *Application of the Inverse*

The cofactor of matrix **A** is given by

$$\text{cof } \mathbf{A} = \begin{bmatrix} +792 & -600 & +984 & -1008 \\ +264 & -200 & -1152 & +1144 \\ -1272 & +2040 & -504 & +408 \\ +552 & -1360 & +1224 & -568 \end{bmatrix}$$

and thus

$$\text{adj } \mathbf{A} = \begin{bmatrix} +792 & +264 & -1272 & +552 \\ -600 & -200 & +2040 & -1360 \\ +984 & -1152 & -504 & +1224 \\ -1008 & +1144 & +408 & -568 \end{bmatrix}$$

Therefore

$$\mathbf{A}^{-1} = \begin{bmatrix} \frac{+792}{8880} & \frac{+264}{8880} & \frac{-1272}{8880} & \frac{+552}{8880} \\ \frac{-600}{8880} & \frac{-200}{8880} & \frac{+2040}{8880} & \frac{-1360}{8880} \\ \frac{+984}{8880} & \frac{-1152}{8880} & \frac{-504}{8880} & \frac{+1224}{8880} \\ \frac{-1008}{8880} & \frac{+1144}{8880} & \frac{+408}{8880} & \frac{-568}{8880} \end{bmatrix}$$

Now,

$$\mathbf{X} = A^{-1}\mathbf{B}$$

Therefore

$$\mathbf{X} = A^{-1} \begin{bmatrix} 1110 \\ 1340 \\ 1320 \\ 1130 \end{bmatrix}$$

and so

$$\begin{bmatrix} x_1 \\ x_2 \\ x_3 \\ x_4 \end{bmatrix} = \begin{bmatrix} +99{\cdot}0 + 39{\cdot}83783 - 189{\cdot}08107 + 70{\cdot}24324 \\ -75{\cdot}0 - 30{\cdot}18018 + 303{\cdot}24324 - 173{\cdot}06306 \\ +123{\cdot}0 - 173{\cdot}83783 - 74{\cdot}91891 + 155{\cdot}75674 \\ -126{\cdot}0 + 172{\cdot}63063 + 60{\cdot}64864 - 72{\cdot}27927 \end{bmatrix}$$

or

$$\begin{bmatrix} x_1 \\ x_2 \\ x_3 \\ x_4 \end{bmatrix} = \begin{bmatrix} 20 \\ 25 \\ 30 \\ 35 \end{bmatrix}$$

Thus again the calorific values of individuals of the four species are calculated as 20 cal, 25 cal, 30 cal, and 35 cal.

Having found the calorific value of one individual of each of the four food species, one can calculate the total calorific values of the species constituting the diets of the fish. The results are shown in Table 9.2.

Table 9.2

	Calorific value of each species			
	Species 1	*Species 2*	*Species 3*	*Species 4*
Fish 1	360	300	240	210
Fish 2	280	200	300	560
Fish 3	160	250	420	490
Fish 4	40	50	480	560

Exercises

1. Given the determinant

$$A = \begin{vmatrix} 4 & 3 & 7 \\ 2 & 4 & 1 \\ 1 & 6 & 5 \end{vmatrix}$$

find the minor and cofactor of the element a_{23}.

2. Calculate the value of det A given in question 1.
3. Solve for x_1, x_2, and x_3 in the equations

$$\begin{aligned} x_1 + 2x_2 + x_3 &= 8 \\ 2x_1 + x_2 + x_3 &= 7 \\ x_1 + x_2 + 3x_3 &= 12 \end{aligned}$$

by the application of Cramer's rule. Check the result by recalculating the unknowns by means of the inverse method.

Ten

REGRESSION ANALYSIS BY MATRICES

10.1. Regression Analysis

The outcome of many ecological studies is that certain reactions of individuals can often be expressed as a function of an independent variable. For example, the author's own research has shown that the degree of activity amongst an aquatic insect population can be directly related to the level of light in the surrounding environment. At the end of such work, one usually has a set of values for the dependent and independent variables, and it would be convenient if the data could be used to develop an equation describing the relationship between the variables.

Such an equation can be achieved by regression analysis, whereby one attempts to fit the data to a straight line with the general formula

$$y = a + bx \qquad (10.1)$$

where

y = any value of the dependent variable,
x = any value of the independent variable, and
a and b are calculated constants.

In the analysis, it is the calculation of a and b which is important because they are the terms which correct for the ranges of the observed values of the variables about their respective means. Thus, if

x = the observed values of the independent variable,
$\bar{x}$ = the mean (i.e. average) value of x,
y = the observed values of the dependent variable, and
$\bar{y}$ = the mean value of y,

then the constants can be calculated because

$$b = \frac{(x - \bar{x})(y - \bar{y})}{(x - \bar{x})^2} \qquad (10.2)$$

and

$$a = \bar{y} - b\bar{x} \qquad (10.3)$$

This method of fitting data to a regression line is known as 'the method of least squares', and for further information on its application, the reader is referred to any introductory text on statistics.

10.2. Biological Example

A study was made on the number of attacks made by a predator, in nine different areas, in relation to the number of prey and the vegetation cover in each area. The vegetation cover was estimated on a points basis, which ranged from zero for no cover up to 15 for total cover. The data from the observations were as shown in Table 10.1 (note that the numbers of prey have been reduced to one-tenth of their actual values to facilitate an easier analysis).

Table 10.1

Area	*Number of attacks,* y_i	*Number of prey,* x_{i1} ($\times 0{\cdot}1$)	*Cover,* x_{i2}
1	16	5	3
2	16	5	6
3	27	5	12
4	18	10	3
5	20	10	6
6	28	10	12
7	26	15	3
8	27	15	6
9	32	15	12

In the above example, the number of attacks is a function of two independent variables, i.e. the number of prey and the vegetation cover. It is still possible to calculate the relationship between the variables using the least-squares method, except that it should be noted that

$$y = a + b_1x_1 + b_2x_2 \tag{10.4}$$

is an equation for a regression plane, as opposed to equation 10.1 which is an equation for a regression line.

The values of the constants a, b_1, and b_2 again correct for the ranges of the data about the means, but the calculation is complicated by the need for the error corrections to express:

1. The deviation of values of the independent variables from their respective means.

2. The pooled deviation of both independent variables from their respective means.
3. The pooled deviation of the independent variables and the dependent variable from their respective means.

Although it is still possible to analyse the data with the least-squares method, the calculation, which is known as multiple regression analysis, is greatly facilitated by the introduction of matrix notation.

The data can be rewritten in the form

$$\mathbf{Y} = \begin{bmatrix} 16 \\ 16 \\ \vdots \\ 27 \\ 32 \end{bmatrix} \qquad \mathbf{X} = \begin{bmatrix} 5 & 3 \\ 5 & 6 \\ \vdots & \vdots \\ 15 & 6 \\ 15 & 12 \end{bmatrix}$$

For the calculation of the error corrections, it is necessary to separate the two independent variables contained in the $\mathbf{X}$ matrix. This can be done by differentiating $\mathbf{X}$ with respect to x_{i1} (the number of prey) and x_{i2} (the vegetation cover), since

$$\left(\frac{d\mathbf{X}}{dx_{in}}\right)_{n=1,2} = \begin{bmatrix} \dfrac{d\mathbf{X}}{dx_{i1}} \\ \dfrac{d\mathbf{X}}{dx_{i2}} \end{bmatrix}$$

where

$$\frac{d\mathbf{X}}{dx_{i1}} = \begin{bmatrix} 5 \\ 5 \\ \vdots \\ 15 \\ 15 \end{bmatrix} \qquad \frac{d\mathbf{X}}{dx_{i2}} = \begin{bmatrix} 3 \\ 6 \\ \vdots \\ 6 \\ 12 \end{bmatrix}$$

It is now possible to calculate

$$\sum_{i=1}^{9} y_i = 210$$

which is found by summing all the terms in matrix $\mathbf{Y}$, and

$$\sum_{i=1}^{9} x_{i1} = 90$$

which is found by summing all the values in $d\mathbf{X}/dx_{i1}$. Therefore,

$$(\sum_{i=1}^{9} x_{i1})^2 = 90^2 = 8100$$

and

$$\sum_{i=1}^{9} (x_{i1})^2 = 1050$$

which is found by multiplying matrix $d\mathbf{X}/dx_{i1}$ by its transpose.
Similarly,

$$\sum_{i=1}^{9} x_{i2} = 63$$

and

$$(\sum_{i=1}^{9} x_{i2})^2 = 63^2 = 3969$$

and

$$\sum_{i=1}^{9} (x_{i2})^2 = 567$$

Also,

$$\sum_{i=1}^{9} (x_{i1}x_{i2}) = 630$$

which can be found by multiplying the transpose of $d\mathbf{X}/dx_{i1}$ by the column matrix $d\mathbf{X}/dx_{i2}$.
The summation

$$\sum_{i=1}^{9} (y_i x_{i1}) = 2230$$

can be found by multiplying the transpose of $\mathbf{Y}$ by $d\mathbf{X}/dx_{i1}$. Similarly

$$\sum_{i=1}^{9} (y_i x_{i2}) = 1602$$

The average

$$\bar{x}_{i1} = \frac{\sum_{i=1}^{9} x_{i1}}{n} = \frac{90}{9} = 10$$

can be found by summing all the values in $d\mathbf{X}/dx_{i1}$ and dividing by the number of observations. Similarly,

$$\bar{x}_{i2} = \frac{\sum_{i=1}^{9} x_{i2}}{n} = \frac{63}{9} = 7$$

and

$$\bar{y}_i = \frac{\sum_{i=1}^{9} y_i}{n} = \frac{210}{9} = 23\tfrac{1}{3}$$

It is now possible to construct a matrix which describes the deviation of the values of the independent variables from their means and the pooled deviation of these variables from the means. The form of this matrix is

$$\mathbf{A} = \begin{bmatrix} \sum_{i=1}^{9} (x_{i1})^2 - \dfrac{(\sum_{i=1}^{9} x_{i1})^2}{n} & \sum_{i=1}^{9} (x_{i1}\, x_{i2}) - \dfrac{\sum_{i=1}^{9} x_{i1} \sum_{i=1}^{9} x_{i2}}{n} \\ \sum_{i=1}^{9} (x_{i1}\, x_{i2}) - \dfrac{\sum_{i=1}^{9} x_{i1} \sum_{i=1}^{9} x_{i2}}{n} & \sum_{i=1}^{9} (x_{i2})^2 - \dfrac{(\sum_{i=1}^{9} x_{i2})^2}{n} \end{bmatrix}$$

$$= \begin{bmatrix} 1050 - \dfrac{8100}{9} & 630 - \dfrac{90 \times 63}{9} \\ 630 - \dfrac{90 \times 63}{9} & 567 - \dfrac{3969}{9} \end{bmatrix}$$

$$= \begin{bmatrix} 150 & 0 \\ 0 & 126 \end{bmatrix}$$

from which one can calculate the cofactor and adjoint of matrix $\mathbf{A}$. Thus, since

$$\det A = 18\,900$$

it follows that

$$\mathbf{A}^{-1} = \begin{bmatrix} \dfrac{126}{18\,900} & 0 \\ 0 & \dfrac{150}{18\,900} \end{bmatrix}$$

It is also possible to construct a matrix which describes the pooled deviation of the independent variables and the dependent variable from the means, such that

$$\mathbf{B} = \begin{bmatrix} \sum_{i=1}^{9} (y_i x_{i1}) - \dfrac{\sum_{i=1}^{9} x_{i1} \sum_{i=1}^{9} y_i}{n} \\ \sum_{i=1}^{9} (y_i x_{i2}) - \dfrac{\sum_{i=1}^{9} x_{i2} \sum_{i=1}^{9} y_i}{n} \end{bmatrix}$$

$$= \begin{bmatrix} 2230 - \dfrac{90 \times 210}{9} \\ 1602 - \dfrac{63 \times 210}{9} \end{bmatrix}$$

$$= \begin{bmatrix} 130 \\ 132 \end{bmatrix}$$

The constants b_1 and b_2 can be calculated from $\mathbf{A}^{-1}$ and $\mathbf{B}$ because

$$\begin{bmatrix} b_1 \\ b_2 \end{bmatrix} = \mathbf{A}^{-1}\mathbf{B}$$

$$= \begin{bmatrix} \dfrac{126}{189\,000} & 0 \\ 0 & \dfrac{150}{189\,000} \end{bmatrix} \begin{bmatrix} 130 \\ 132 \end{bmatrix}$$

$$= \begin{bmatrix} \dfrac{16\,380}{189\,000} & 0 \\ 0 & \dfrac{19\,800}{189\,000} \end{bmatrix}$$

and therefore $b_1 = 0{\cdot}867$ and $b_2 = 1{\cdot}05$.

Finally, constant a can be calculated because

$$a = \bar{y} - (b_1\bar{x}_{i1} + b_2\bar{x}_{i2})$$

$$= \frac{210}{9} - (0{\cdot}867 \times 10) - (1{\cdot}05 \times 7)$$

$$= 7{\cdot}33$$

Thus the estimation of the regression plane is

$$y = a + b_1x_1 + b_2x_2$$

$$= 7{\cdot}33 + 0{\cdot}867x_1 + 1{\cdot}05x_2$$

This example is a simple case of only two independent variables and nine observations. In more complex situations, the approach is exactly the same except that matrices **A** and **B** are enlarged. If the reader is not convinced that the use of matrices simplifies multivariate analysis, it is suggested that he reads the excellent text *Statistical Methods in Biology* by N. T. J. Bailey published by The English Universities Press (1959), to see how complex the least-squares approach to multivariate analysis can quickly become without the recourse to matrix notation.

Eleven

AN INTRODUCTION TO SOME OF THE BASIC TECHNIQUES OF OPERATIONS RESEARCH

11.1. Introduction

Operations research has evolved to meet the increasing demands from commerce, the Armed Services, and Government departments for accurate predictions of the outcome of their activities and for models to analyse complex situations. This new science utilises various mathematical techniques, and it is probable that it would not have attained its present great success without the assistance of high-speed computers.

The author first came in contact with the science during a year spent at the University of Georgia, USA, where Professor B. C. Patten had recently introduced a postgraduate course on systems analysis. From the initial stimulation of this course, the author has continued to study certain aspects of the science because the texts on the subject often contain excellent introductory chapters on advanced mathematics and in addition describe techniques which may have potential applications in the analysis of biological situations.

This chapter contains an introduction to four methods which are frequently utilised in operations research. Very little mathematical theory is included with the examples because the purpose of this final chapter is to introduce the reader to the potentials of the science for the analysis of biological situations. It is hoped that the next few pages will stimulate the reader into further study of the techniques and perhaps into finding some applications of the science to his own work in biology.

11.2. The Simplex Method of Linear Programming

A study of a predator revealed that it has a nest at a site A and has two potential sources of food, i.e. species x_1 and x_2, which are located in areas B and C respectively (Figure 11.1). The time taken by the predator to travel to area B and return with a single capture was found to be 20 min, and the corresponding travel time for site C was found to be

30 min. In addition, the study revealed the constraint that the animal was not prepared to spend more than 120 min per day in travelling to and from either of the two sites.

It was found that the predator took 20 min to capture a unit of species x_1 in area B and that it took 10 min to capture a unit of x_2 in

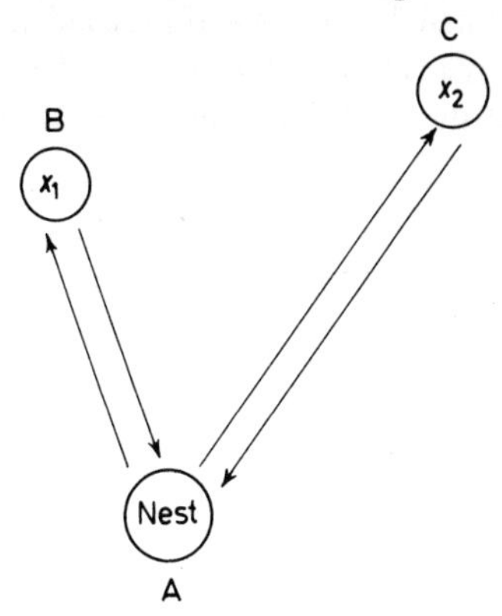

Figure 11.1

area C. In addition, it was found that the predator would not spend more than 80 min per day in searching for prey.

The final data obtained from the study was that the calorific value of one unit of x_1 was 60 cal, and the calorific value of x_2 was 70 cal.

It is now desired to use these data to find how many individuals of species x_1 and x_2 should be captured per day in order that the predator can maximise the calorific return on its hunting activities.

The constraints for the model are: the travelling-time constraint

$$20x_1 + 30x_2 \leqslant 120 \qquad (11.1)$$

the searching constraint

$$20x_1 + 10x_2 \leqslant 80 \qquad (11.2)$$

and the prey species

$$x_1 \geqslant 0 \qquad (11.3)$$

$$x_2 \geqslant 0 \qquad (11.4)$$

i.e. the predator cannot capture a negative amount of either species.

Equation 11.1 causes a constraint whereby the predator can obtain 0 units of x_1 and 4 units of x_2 in the permitted time, or 6 units of x_1 and 0 units of x_2, or any other combination of prey species as long as it is a combination which is possible under the constraint. Further, equation 11.2 causes the constraint whereby the predator can obtain 0 units of x_1 and 8 units of x_2 in the permitted search time, or 4 units of x_1 and 0 units of x_2, or any other combination of prey species as long as it is a combination which is possible under the constraint. When these two constraints are operating together, the combination is even more limited, and the possible numbers of prey are described by area OABC in Figure 11.2. Area OABC is known as the feasible solution of the problem. Thus, if the predator wishes to maximise the calorific return

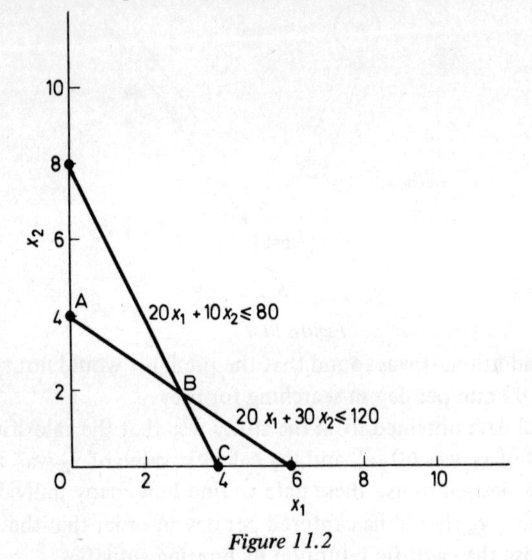

Figure 11.2

from its hunting activities, it should obtain the combination of prey described by point B of Figure 11.2.

If Figure 11.2 were drawn very carefully, it would be possible to read off the combination for maximum calorific return, but, if there are more than two prey species, then the construction of the relevant multi-axis graph would become very difficult. One technique, the simplex method, has thus been developed to overcome this problem by using matrix algebra to calculate the solution without recourse to graphical methods. The simplex method can be used to solve the simple predator problem in this section.

First, it is necessary to introduce two slack variables x_3 and x_4 to convert the inequalities of the constraint equations to equalities. Then equation 11.1 becomes

$$20x_1 + 30x_2 + x_3 + 0x_4 = 120 \tag{11.5}$$

and equation 11.2 becomes

$$20x_1 + 10x_2 + 0x_3 + x_4 = 80 \qquad (11.6)$$

In addition, if

E = the total calorific value of the captured prey,

then from the calorific values of units of x_1 and x_2,

$$E = 60x_1 + 70x_2 + 0x_3 + 0x_4 \qquad (11.7)$$

These equations can be used to construct submatrices which can all be gathered into an augmented matrix $\mathbf{A}$. Thus, from equations 11.5 and 11.6, if

$$\mathbf{B} = \begin{bmatrix} 20 & 30 \\ 20 & 10 \end{bmatrix}, \quad \mathbf{I}_m = \begin{bmatrix} 1 & 0 \\ 0 & 1 \end{bmatrix} \quad \mathbf{C} = \begin{bmatrix} 120 \\ 80 \end{bmatrix}$$

and, from equation 11.7, if

$$\mathbf{D} = \begin{bmatrix} 60 & 70 \end{bmatrix} \quad \mathbf{F} = \begin{bmatrix} 0 & 0 \end{bmatrix} \quad \mathbf{E} = \begin{bmatrix} E \end{bmatrix}$$

then

$$\mathbf{A} = \begin{bmatrix} \mathbf{B} & \mathbf{I}_m & \mathbf{C} \\ \mathbf{D} & \mathbf{F} & \mathbf{E} \end{bmatrix}$$

$$= \begin{bmatrix} 20 & 30 & 1 & 0 & 120 \\ 20 & 10 & 0 & 1 & 80 \\ 60 & 70 & 0 & 0 & E \end{bmatrix}$$

From this augmented matrix, it is possible by a sequence of operations to find the number of units of x_1 and x_2 which the predator should obtain in order to maximise the calorific value of its hunting activities.

1. One purpose of the operations is to make all the entries of the last row of $\mathbf{A}$ non-positive, and to achieve this it is necessary to choose any column (except the last) whose last entry is positive. This latter demand will naturally exclude column 3 and 4 in $\mathbf{A}$ because the entries are zero.
2. A pivot entry must be found by dividing each non-zero element (except the last) of the chosen column into its corresponding entry in the last column. The purpose of this operation is to keep all the entries of the last column non-negative.

Thus, carrying out operations 1 and 2 on $\mathbf{A}$, take column 1 and form the ratios 120/20 and 80/20. As the latter ratio is the smallest, then a_{21} is the pivot entry.

3. All the other entries of the chosen column must be made equal to zero by adding multiples of the entries of the pivot row to the corresponding entries in the other rows.

Thus, carrying out operation 3 on $\mathbf{A}$ using pivot a_{21}, calculate the sums

$$R_1 + (-1)R_2$$

and

$$R_3 + (-3)R_2$$

which yields

$$\mathbf{A} = \begin{bmatrix} 0 & 20 & 1 & 0 & 40 \\ 20 & 10 & 0 & 1 & 80 \\ 0 & 40 & 0 & -3 & E-240 \end{bmatrix}$$

4. Operations 1 – 3 must be repeated on the other columns (except the last) whose last entry is positive until all the entries of the last row are non-positive.

Thus, take the new column 2 and form the ratios 40/20 and 80/10, and, as the former ratio is the smaller, then a_{12} is the pivot entry. To make all the other entries of column 2 equal zero, calculate the sums

$$R_2 + (-\tfrac{1}{2})R_1$$

and

$$R_3 + (-2)R_1$$

which yields

$$\mathbf{A} = \begin{bmatrix} 0 & 20 & 1 & 0 & 40 \\ 20 & 0 & -\frac{1}{2} & 1 & 60 \\ 0 & 0 & -2 & -3 & E-320 \end{bmatrix}$$

As all the entries in the last row are now non-positive, one can pass on to the next operation.

5. A submatrix $\mathbf{I}_m$ must be created, in part of the space formally occupied by $\mathbf{B}$, by dividing the necessary rows by a chosen scalar.

Thus, dividing row 1 by 20 and row 2 by 20 yields

$$\mathbf{A} = \begin{bmatrix} 0 & 1 & \frac{1}{20} & 0 & 2 \\ 1 & 0 & -\frac{1}{40} & \frac{1}{20} & 3 \\ 0 & 0 & -2 & -3 & E-320 \end{bmatrix}$$

From the above operations there are now two columns each with an element of value 1 and with all the other entries of value zero. If the ith column is of this type, then x_i is equal to the last entry of the same row as the 1. If the ith column is not such a column, i.e. with an entry of 1 and all other elements equal to zero, then $x_i = 0$.

Thus, from the above array,

$$x_1 = 3$$

and

$$x_2 = 2$$

Finally, setting the bottom right-hand corner to zero yields

$$E - 320 = 0$$

or

$$E = 320$$

It can therefore be concluded that the predator will obtain a maximum calorific yield of 320 calories from its hunting activities if it captures 3 units of x_1 and 2 units of x_2. The result can be checked from equation 11.7:

$$\begin{aligned} E &= 60x_1 + 70x_2 + 0x_3 + 0x_4 \\ &= (60 \times 3) + (70 \times 2) + 0 + 0 \\ &= 320 \end{aligned}$$

which corresponds to the calculated maximum in the matrix operations.

When examining the maximum problem, it should be noted that there exists a 'dual' such that it is possible to use the transpose of $\mathbf{A}$ to calculate the minimum values of x_i to be collected to satisfy a given value of E. Thus the simplex method can be used to calculate both minimum and maximum values for variables which are subject to constraints.

If the reader ever has to apply the simplex method, he is warned that he may not necessarily obtain a unique solution for the problem. The usual cause of this situation can be attributed to a mis-stated constraint, and therefore the initial conditions of the system should be re-examined in an attempt to trace the fault.

11.3. The Transportation Method

Four individuals of the same species have nests containing young, and there are four possible areas from which the individuals can obtain food for their young (Figure 11.3). The number of food units required by the individuals (N_i) and the potential supplies S_i (where $i = 1, 2, 3, 4$) in the four areas are as shown in Table 11.1. The energetic costs (in

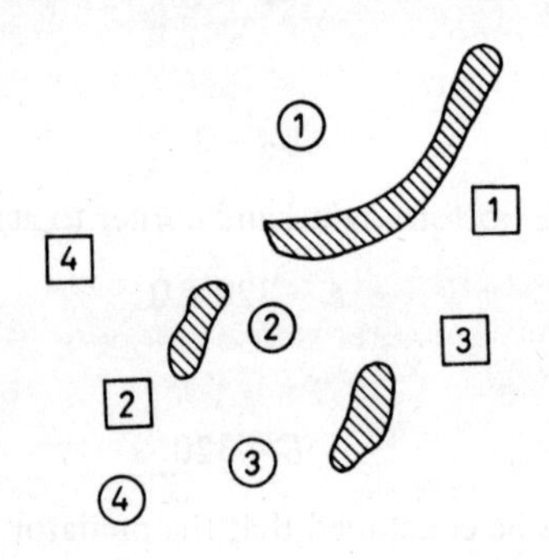

Figure 11.3

Table 11.1

Requirements		*Supplies*	
N_1	10	S_1	25
N_2	20	S_2	10
N_3	15	S_3	10
N_4	15	S_4	15

calories) incurred by the respective individuals in obtaining one unit of food from the various areas are

S_i / N_i	S_1	S_2	S_3	S_4
N_1	26	20	28	18
N_2	14	15	16	21
N_3	19	16	20	25
N_4	30	13	28	24

These variations of cost can be attributed to the location of the food supplies in relation to the four nests and to the varying efficiencies of the four individuals in obtaining their food.

This type of data can be analysed to see if the total energy expended by the four individuals has been minimised. The analytical approach involves the application of the transportation method, which is a technique that has been developed for solving costing strategies for firms involved in distributing their products to industrial and consumer markets.

The first step in the analysis involves finding a feasible solution whereby all the individuals can satisfy their demands for food. This is achieved by a sequence of operations known as the northwest corner technique, whereby the needs of each individual are compared with the supplies available.

First, a table with supplies along the top and needs down the left-hand side is set up. Then, starting at the northwest corner, the need is compared with supply:

1. If $S_1 > N_1$, set $x_{11} = N_1$, modify $S_1' = S_1 - N_1$, and move vertically down the table.
2. If $S_1 = N_1$, set $x_{11} = N_1$ and move diagonally to the next cell (x_{22}).
3. If $S_1 < N_1$, set $x_{11} = S_1$, modify $N_1' = N_1 - S_1$, and proceed to the adjacent cell (x_{12}).

This operation is continued until a value is reached in the southeast corner.

Use of this approach on the data for supply and need yields

N_i \ S_i	25	10	10	15
10	10 ↓			
20	15 →	5 ↓		
15		5 →	10 ↘	
15				15

This shows a feasible solution for obtaining food, whereby all four individuals have satisfied their requirements.

The next step is to find the strategy which will minimise the total food-obtaining costs and then to compare the result with the feasible solution to find if there is a difference in costs. The minimisation of costs involves a number of operations.

First, the costs data are inserted into a table which has the same border values as those used in the feasible solution. Thus

N_i \ S_i	25	10	10	15
10	26	20	28	18
20	14	15	16	21
15	19	16	20	25
15	30	13	28	24

The minimum value in each column is found and subtracted from all the entries in that column. This process gives

N_i \ S_i	25	10	10	15
10	12	7	12	0
20	0	2	0	3
15	5	3	4	7
15	16	0	12	6

The position of each zero is noted, then for each zero the supply for its column is placed in brackets above the corresponding row need. The converse operation, such that the row need is inserted in brackets above the corresponding supply, is also carried out. Thus

N_i \ S_i	(20) 25	(15) 10	(20) 10	(10) 15
(15) 10	12	7	12	0
(25 + 10) 20	0	2	0	3
(0) 15	5	3	4	7
(10) 15	16	0	12	6

Next, where the value in brackets exceeds or equals the value below it, a line is drawn connecting the entries in the corresponding row or column:

N_i \ S_i	(20) 25	(15) 10	(20) 10	(10) 15
(15) 10	---12---	---7---	---12---	---0---
(25 + 10) 20	---0---	---2---	---0---	---3---
(0) 15	5	3	4	7
(10) 15	16	0	12	6

If any border value is greater than the bracket entry directly above it, then the smallest value of the table entries not connected by any line must be subtracted from any remaining entries not connected by any line. An additional zero is created, and therefore the supply value for that column is added to the value in the bracket for the corresponding row. This operation is repeated with the row need, which is added to the value in the bracket entry for the corresponding column supply. Thus

N_i \ S_i	(20 + 15) 25	(15) 10	(20) 10	(10) 15
(15) 10	12	7	12	0
(25 + 10) 20	0	2	0	3
(25) 15	0	3	4	2
(10) 15	11	0	12	1

This last operation is repeated until all the border values are less than or equal to their bracket entries. When this latter point is reached, one has obtained the optimal solution, which is

N_i \ S_i	(20 + 15) 25	(15) 10	(20) 10	(10 + 15) 15
(15) 10	12	7	12	0
(25 + 10) 20	0	2	0	3
(25) 15	0	3	4	1
(10 + 15) 15	10	0	12	0

Now any row of the optimal solution which contains a single zero entry is taken and that entry is given a maximum supply:

N_i \ S_i	25	10	10	15
10				10
20				
15	15			
15				

The remaining supplies are then allocated to the rows which have more than one zero. Finally, any remaining supplies are allocated to individuals who have not satisfied their needs. These supply allocations can then be used to compute the total cost incurred by the individuals while obtaining their food. Thus the food allocations are

N_i \ S_i	25	10	10	15
10				10
20	10		10	
15	15			
15		10		5

and total costs are

$$
\begin{aligned}
10 \times 18 &= 180 \text{ cal} \\
10 \times 14 &= 140 \text{ cal} \\
10 \times 16 &= 160 \text{ cal} \\
15 \times 19 &= 285 \text{ cal} \\
10 \times 13 &= 130 \text{ cal} \\
5 \times 24 &= 120 \text{ cal} \\
\hline
& 1015 \text{ cal}
\end{aligned}
$$

Finally, the total costs using the feasible solution are calculated. If the feasible solution incurs a greater cost, then the allocation of supplies should be done using the minimum cost solution in order that the total food-obtaining costs be at their lowest possible level. Costs of the feasible solution are

$$
\begin{aligned}
10 \times 26 &= 260 \text{ cal} \\
15 \times 14 &= 210 \text{ cal} \\
5 \times 15 &= 75 \text{ cal} \\
5 \times 16 &= 80 \text{ cal} \\
10 \times 20 &= 200 \text{ cal} \\
15 \times 24 &= 360 \text{ cal} \\
\hline
& 1185 \text{ cal}
\end{aligned}
$$

In this case, there would be a marginal decrease in costs if the minimum cost strategy were used in place of the feasible solution.

11.4. Critical Path Analysis

Critical path analysis has evolved to satisfy the need for a method of analysing long-term projects which involve many interlinked activities. By describing the project as a network of events, it is possible to evaluate the activities which take the longest time; this evaluation is important because it will be these activities which decide the shortest time in which the project can be completed, i.e. they constitute the critical paths of the network.

The first two major applications of critical path analysis were PERT (Programme Evaluation and Review Techniques), which was used in the 1958 Polaris missile programme, and CPM (Critical Path Method), which was first used by the Du Pont de Nemours Company in 1957. Since this time, there has been a tremendous proliferation of systems based upon the network concept, and two of the most well-known acronyms for this type of work are PORT (Project Orientation and Rescheduling Technique) and BEAUJOLAIS (Batch Evaluation of Aircraft Units for the Joint Orientation of Line Assembly Integrated Systems).

The basic principles of networks can be illustrated by studying a simplified version of the innate behaviour pattern of the male stickleback. There are a number of parallel activities which have to be completed before egg laying can occur in this species. First, increasing day length in the Spring stimulates the animal and results in the maturation of hormonal centres controlling: (a) the drive to establish a territory and build a nest, (b) maturation of the gonads, and (c) the deposition of epidermal pigments which are necessary to stimulate the female. The hormones released from the mature centres cause the male to carry out all the activities that are necessary before the mating and egg laying can occur.

The process can be illustrated by a network, where

O symbolises event

⟶ symbolises interconnecting activities

---> symbolises instantaneous activities

and the time taken for each activity is inserted above the activity. In Figure 11.4, this approach is applied to the innate reproductive behaviour of the male stickleback. Note that the activity times are hypothetical and have only been adopted for this specific example.

The network can be used to find the critical path, but it is first necessary to carry out a notation operation because the events have to be numbered in the order in which they occur. Thus, starting at the beginning of the network, one can label 'day length' as event 0. The next event that is completed is the maturation of the hormonal centre controlling pigmentation, and thus this is labelled as event 1. This numbering is continued until all the events have been labelled in their order of occurrence (Figure 11.5).

The longest path can be found by starting at the final event and summing all the activity times along each path. Thus

Colouration path = 4 + 1 = 5 weeks
Gonadial development = 4 + 2 = 6 weeks
Territory and nest = 1 + 3 + 4 = 8 weeks

It can therefore be concluded that the territorial and nest construction path is the longest, i.e. the critical path, because it takes 8 weeks. In addition, there is 'slack' time in the other two paths because they are completed in 5 weeks and 6 weeks.

Once the critical path has been identified, it is possible to manipulate the times along the critical path to see if any reduction in the activity times can shorten the completion time for the network. For example, if maturation of the hormonal centre is reduced by 1 week, the critical time becomes

$$1 + 3 + 3 = 7 \text{ weeks}$$

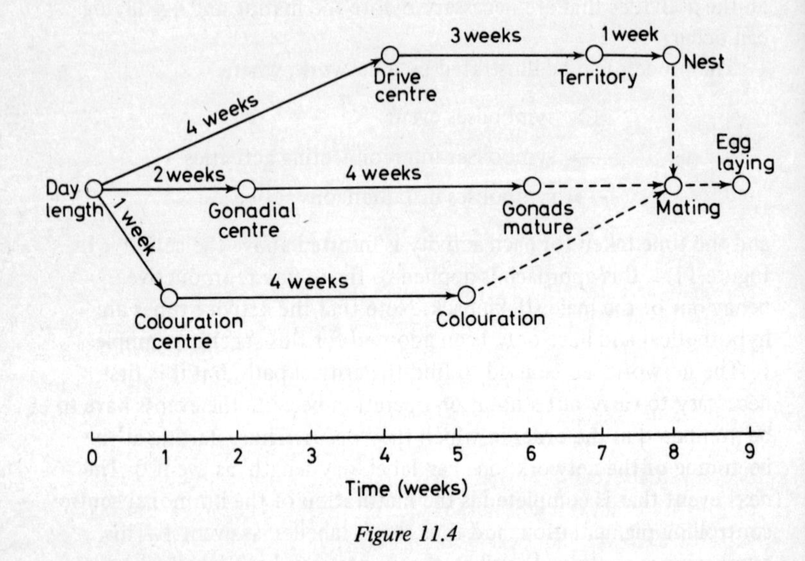

Figure 11.4

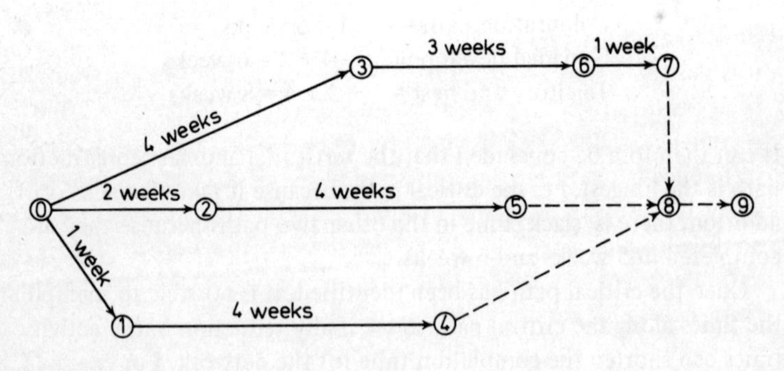

Figure 11.5

Thus the completion date is reduced to 7 weeks although this same path would still be the critical path.

If the maturation is reduced by another week, then the critical path time becomes

$$1 + 3 + 2 = 6 \text{ weeks}$$

This reduces the completion time to 6 weeks and means that this path and the gonadial development path are then both critical paths.

If the hormonal maturation for controlling the territorial drive is reduced by a further week, then the time of this path becomes

$$1 + 3 + 1 = 5 \text{ weeks}$$

and, if this occurs, the gonadial maturation becomes the critical path.

The usual reason for manipulation of this type on a network is to see if resources can be transferred from shorter activities and thus reduce the completion time. In the above example, the manipulation is quite simple, but one can imagine the immense problems in carrying out such iterations by hand on a large network. The digital computer then becomes a useful aid in that it has unique abilities in solving such a problem. In fact, project networks are usually so complex today that very few are ever analysed by hand; the majority are immediately put into the standard network programmes that are available at most computer centres.

Often, networks which have more than one value for each activity are encountered. For example, the construction industry usually utilises networks in which each activity is given a pessimistic time, a most likely time, and an optimistic time. With this type of information, it is possible to incorporate probability theory such that the completion time for the project is given as a value plus or minus the calculated confidence limits. The magnitude of these limits is decided by the difference between the pessimistic and optimistic times. If network analysis is ever widely applied to biological data, it is probable that the confidence limit approach will be needed to compensate for the extreme ranges of variables that one usually encounters whilst accumulating ecological data.

11.5. Queueing Theory

The life of socialised insects with division of labour and large colonies is an aspect of entomology which easily lends itself to analysis in terms of the economics of the community's activities. For example, the older honeybees occupy their day by leaving the nest and returning with nectar and pollen that they have obtained from flowers. Upon the arrival of these honeybees at the nest, it is usual for a younger worker to unload the nectar and take it into the nest for processing. Naturally, if there were only one worker unloading the incoming bees, a queue would soon form and the collecting bees would be wasting valuable time which could be better spent collecting more food.

The economics of this situation can be analysed by applying queueing theory, which is a technique that has been developed to assess the time spent waiting by units wishing to be serviced. A typical example is the analysis of people waiting to pay for their goods at the check-out desk of a supermarket. Obviously, the time spent in waiting by any one customer is dependent upon the number of customers arriving at the check-out and upon the time taken for the supermarket employee to estimate the value of the purchases of each customer.

The honeybee problem can be placed on a more formal basis if

B = average number of bees returning to the nest with food per hour,
W = average number of bees that the worker can unload per hour,
m = number of bees being unloaded or waiting to be unloaded,
n = number in the queue (not including the bee actually being unloaded),
$E(m)$ = expected number in the queue or being unloaded,
$E(n)$ = expected number in the queue, and
$E(t)$ = average waiting time in the queue.

If $B/W < 1$, then

$$E(m) = \frac{B}{W-B} \tag{11.7}$$

and

$$E(n) = \frac{B^2}{(W-B)W} \tag{11.8}$$

and

$$E(t) = \frac{E(n)}{B} \tag{11.9}$$

For example, if the number of bees arriving per hour is 25 and the number of bees unloaded by the worker per hour is 30, then

$$\frac{B}{W} = \frac{25}{30}$$

which is less than 1. So, from equation 11.7, the expected number in the queue or being unloaded is given by

$$E(m) = \frac{25}{30 - 25} = 5$$

From equation 11.8, the average number in the queue is

$$E(n) = \frac{25^2}{(30 - 25)30} = \frac{25}{6}$$

and thus, from equation 11.9, the waiting time for an arriving bee is

$$E(t) = \frac{25}{6 \times 25} = \frac{1}{6}$$

Thus the waiting time for the bee is $\frac{1}{6}$ hours = 10 min.

The total amount of time T (in hours) wasted by waiting during 12 hour working day is given by

$$\begin{aligned} T &= 12\,BE(t) \\ &= 12 \times 25 \times \frac{1}{6} \\ &= 50 \end{aligned}$$

and, if a bee collects 1 g of nectar per hour, then the potential loss in nectar for the hive is 50 g of nectar per day.

Under normal circumstances, one would expect there to be more than one worker unloading the nectar. In this situation, if

p_0 = probability of having no bees arriving or being unloaded, and
N = number of workers unloading,

then

$$p_0 = \left[\frac{(B/W)^N}{N!(1 - B/WN)} + 1 + \frac{(B/W)^1}{1!} + \frac{(B/W)^2}{2!} + \ldots + \frac{(B/W)^{N-1}}{(N-1)!}\right]^{-1}$$

where $N!$ denotes N factorial (e.g. $3! = 3 \times 2 \times 1 = 6$). With N workers unloading, the average number of bees in the queue is

$$E(n) = \frac{(B/W)^{N+1}}{N(N!)(1 - B/WN)^2}$$

For example, if $B = 5$, $W = 10$, and $N = 3$, then $B/W = 0{\cdot}5$ and

$$p_0 = \frac{1}{\dfrac{(0{\cdot}5)^3}{3!(1 - 0{\cdot}5/3)} + 1 + \dfrac{(0{\cdot}5)^1}{1!} + \dfrac{(0{\cdot}5)^2}{2!}}$$

$$= \frac{1}{0{\cdot}025 + 1 + 0{\cdot}5 + 0{\cdot}125}$$

$$= 0{\cdot}6$$

Thus

$$E(n) = \frac{(0{\cdot}5)^4 \times 0{\cdot}6}{3(3!)(1 - 0{\cdot}5/3)^2} = 0{\cdot}003$$

Therefore the average number of bees in the queue is 0·003, and the expected waiting time (from equation 11.9) is

$$E(t) = \frac{0{\cdot}003}{5} = 0{\cdot}0006$$

This results in the total waiting time during a 12 hour day of

$$12 \times 5 \times 0{\cdot}0006 = 0{\cdot}036 \text{ hours}$$

Naturally, this result is a great improvement on the waiting time when there is only one worker unloading the incoming honeybees.

In both the above examples, it has been assumed that there is an average number of bees arriving per hour and that these arrivals occur at regularly spaced intervals. Under normal conditions, one would not expect such a regular arrival pattern, and thus the basic model for queueing theory does not simulate the true conditions of the system. To produce a more natural simulation, it is possible to examine the system with the number of arrivals generated as a random process by means of the Monte Carlo technique. By this method, any arrival frequency is assigned a probability value. Random numbers can be generated from a computer, and each arrival frequency is given a range of random numbers. In this way, the efficiency of the service installation can be tested for any arrival probability.

BIBLIOGRAPHY

This bibliography has been divided into three sections: the first lists formal textbooks which are sources of further mathematical theory; the second deals with biological textbooks concerned with introducing biologists to mathematical theory; and the third contains texts in which ecological theories are discussed in mathematical terms.

Mathematical Textbooks

CAMPBELL, H. G., *An Introduction to Matrices, Vectors and Linear Programming,* Appleton-Century-Crofts, New York (1965)

LANG, S., *A First Course in Calculus,* Addison-Wesley, Reading, Mass. (1968)

MAXWELL, E. A., *An Analytical Calculus* (4 vols), Cambridge University Press, Cambridge (1954)

MORONEY, M. J., *Facts from Figures,* Penguin, Harmondsworth (1962)

PIAGGO, H. T. H., *An Elementary Treatise on Differential Equations and Their Application,* Oliver & Boyd, Edinburgh (1946)

Biomathematical Textbooks

BAILEY, N. T. J., *Statistical Methods in Biology,* English Universities Press, London (1959)

MAYNARD SMITH, J., *Mathematical Ideas in Biology*, Cambridge University Press, Cambridge (1968)

SEARLE, S. R., *Matrix Algebra for the Biological Sciences*, Wiley, New York (1966)

SMITH, C. A. B., *Biomathematics*, Griffin, London (1954)

Ecological Textbooks

HAZEN, W. E., *Readings in Population and Community Ecology*, Saunders, Philadelphia (1965)

LEVINS, R., *Evolution in Changing Environments*, Princeton University Press, Princeton (1968)

LOTKA, A. J., *Elements of Mathematical Biology*. Dover Publications, New York (1956)

MACARTHUR, R. H., and WILSON, E. O., *The Theory of Island Biogeography*, Princeton University Press, Princeton (1967)

SLOBODKIN, L. B., *Growth and Regulation of Animal Populations*, Holt, Rhinehart & Winston, New York (1961)

WATT, K. E. F., *Systems Analysis in Ecology*, Academic Press, New York (1966)

WATT, K. E. F., *Ecology and Resource Management – A Quantitative Approach*, McGraw-Hill, New York (1968)

ANSWERS TO EXERCISES

Chapter 1

1. Length = 36 cm

Chapter 2

1. $2x$
2. $3x^2$
3. $10x^4$
4. 0
5. $3 + 3x^2$
6. $6x^2$
7. $(6x - 4x^2)/x^4$
8. $\exp x$
9. $2 \exp 2x$
10. $3x^2 \exp x^3$

Chapter 4

1. $f_{x_1} = 2x_1x_2^3$; $f_{x_2} = 3x_1^2x_2^2$; $f_{x_3} = 4x_3^3y_1^2$; $f_{y_1} = 2x_3^4y_1$; $f_{y_2} = 3y_2^2y_3^2$; $f_{y_3} = 2y_2^3y_3$
2. $f_x = 2xy^3$; $f_y = 3x^2y^2$; $f_{xx} = 2y^3$; $f_{yy} = 6x^2y$; $f_{xy} = 6xy^2$; $f_{yx} = 6xy^2$
3. A maximum at $x = 3, y = 4$
4. An extreme point at $x = 2, y = 4$; thus $8xy$ has an extreme value of 64

Chapter 5

1. $\frac{2}{3}x^3 + C$
2. $x^5 + C$
3. $\frac{3}{4}x^{4/3} + C$
4. $(\exp x) + C$

5. $\frac{1}{3}(\exp 3x) + C$
6. $\frac{1}{3}(x^4 + 4)^3 + C$
7. 21

Chapter 8

1. $\begin{bmatrix} 5 & 3 & 8 \\ 5 & 8 & 6 \end{bmatrix}$

2. $\begin{bmatrix} 2 & 6 \\ 6 & 1 \end{bmatrix}$

3. $\begin{bmatrix} 33 & 10 \\ 18 & 6 \end{bmatrix}$

4. $$\begin{bmatrix} \sum_{x=1}^{2} a_{1x}b_{x1} & \sum_{x=1}^{2} a_{1x}b_{x2} \\ \sum_{x=1}^{2} a_{2x}b_{x1} & \sum_{x=1}^{2} a_{2x}b_{x2} \end{bmatrix} = \left[\sum_{x=1}^{2} a_{ix}b_{xj}\right]_{(i=1,2;\ j=1,2)}$$

Chapter 9

1. Minor of a_{23} is $\begin{vmatrix} 4 & 3 \\ 1 & 6 \end{vmatrix}$; cofactor of a_{23} is -21
2. 85
3. $x_1 = 1,\ x_2 = 2,\ x_3 = 3$

INDEX